P9-DOA-541

One Wild Bird at a Time

One
Wild Bird
at a Time

Portraits of Individual Lives

——— ≋ ———

BERND HEINRICH

Houghton Mifflin Harcourt
Boston   New York
2016

MORRILL MEMORIAL LIBRARY
NORWOOD, MASS 02062

598
Heinrich

Text and illustrations copyright © 2016 by Bernd Heinrich

All rights reserved

For information about permission to reproduce selections from
this book, write to trade.permissions@hmhco.com or to
Permissions, Houghton Mifflin Harcourt Publishing Company,
3 Park Avenue, 19th Floor, New York, New York 10016.

www.hmhco.com

Library of Congress Cataloging-in-Publication Data
Names: Heinrich, Bernd, date, author.
Title: One wild bird at a time : portraits of individual lives /
Bernd Heinrich.
Description: Boston : Houghton Mifflin Harcourt, 2016. | Includes index.
Identifiers: LCCN 2015037246 |
ISBN 9780544387638 (hardcover) | ISBN 9780544386402 (ebook)
Subjects: LCSH: Birds — Maine — Anecdotes.
Classification: LCC QL684.M2 H45 2016 | DDC 598 — dc23
LC record available at http://lccn.loc.gov/2015037246

Book design by Lisa Diercks

Printed in the United States of America
DOC 10 9 8 7 6 5 4 3 2 1

Portions of Bernd Heinrich's "Chickadees in Winter"
first appeared in *Natural History Magazine*.

Contents

——

Introduction

—

As a child I had Jacob, a pet crow. I roamed the woods hunting his food — frogs, field mice, caterpillars, beetles, and grasshoppers — and feeding him by hand. But wild birds remained out of reach. Much later my vocation and my passion led me to try to answer specific questions about what animals do and why. The protocol for my investigations required eliminating both random chance and individual differences. However, both of these are important parts of life, not peripheral to it, and the goal of biology is to understand life in nature. In this book I hope to celebrate individuals as they presented themselves during my encounters with them in the wild.

My quest to understand what could be gained from intimacy with wild birds was kindled by an incident in Africa.

Along a dusty dirt road on a hillside in Kenya, I viewed Lake Nakuru with its million pink flamingos and thousands of white pelicans. It was a scene and a palette, beautiful but remote, that made me aware of myself as a tourist in alien territory. But then with a jolt I saw a glimpse of another Eden in the nearby bushland.

Among acacias near the lakeshore, birds were mingling with

the most fearsome mammals on Earth, the powerful and aggressive Cape buffalo. Oxpeckers (a species of starlings) were riding on their backs, while cattle egrets accompanied them underfoot. Blue rollers swirled over their backs hawking insects, and emerald-green sunbirds foraged from red flowers close by. Most wild birds stay separate from humans, yet here many acted as if the buffalo were just part of the scenery. For a moment, putting humans in the place of the Cape buffalo, I wondered: What if all birds treated us like this, or as Jacob had treated me? One of the reasons the world can be exciting and beautiful for us is that perhaps we alone have the capacity to enter vicariously into the worlds of others through knowledge leading to empathy. When getting to know a bird — by learning where it lives, what it eats, how it forages, where and how it nests, what it fears, and in general what it likes and dislikes — we are entering another world. Each animal gives us a new view, a new experience, that involves stepping out of our own world into another, and it is always an adventure.

Birds brighten our days by their otherworldly songs, their visual beauty, and their astonishing behaviors when they are able to be themselves in their natural environments. Our usual distance from them makes it difficult for us to identify and isolate individuals, much less keep track of them in the wild where they conduct their lives. Prying meaning and intimacy from them usually requires long-term and technically difficult studies. Traditionally these have involved attaching leg rings or wing tags of different colors and numbers, or monitoring individuals with electronic devices. Most of these methods are inaccessible to amateur biologists, and in fact only a few professional ornithologists use them. Aside from certain obvious exceptions, the observations discussed in this book were of wild birds unencumbered by devices; I describe a relationship with wild birds that is available to almost anyone anywhere.

My watching birds of a particular species was usually stimulated by an anomalous observation that sparked a question. I followed leads as they presented themselves, waiting for dots to connect and make an interesting pattern or a tentative hypothesis, which then might lead to a tentative revelation. The leads, the twists and turns, made adventures that I have tried here to preserve in written and sometimes literal sketches.

Most of the material for this book was gathered in or near a clearing in the Maine woods next to the cabin where I now live. Large windows in all directions make this home a live-in bird blind. My outhouse with an almost equally panoramic view serves as one as well. The clearing where the cabin stands, an island in the midst of forest, has an abundance of berries, seeds, and insects not found in the closely surrounding woods. As a permanent resident with no radio, television, or other electronic distractions except e-mail, I engage with my avian neighbors, visitors, and vagrants, and keep daily records throughout spring, summer, fall, and winter.

I hope in *One Wild Bird at a Time* to reveal details of birds' everyday lives as seen by direct observation, and while I hope also to capture something of the adventure of the chase, this book is less about research results than about the reasons I "do" research. It is aimed to be as realistic as science demands and imaginative enough to suggest possibilities that science allows.

One Wild Bird at a Time

—————≋—————

Flickers in the House

MY SUMMER SPENT OBSERVING BIRDS SEEMED TO BE WIND-
ing down. Most of them had found mates, made nests, and
incubated their eggs; feeding the young was now their main pre-
occupation. I had finished a marathon observation of a pair of
tree swallows. My phoebe had this year not attracted a mate. The
blue-headed vireos' story was finished for the year: one nest I was
watching in a balsam fir tree had fledged four young; the other
had been abandoned after a gradual decline in the number of eggs
in the nest. The sapsuckers with their "super drum" on the apple
tree had stopped drumming and were now incubating or feeding
young. I was starting to feel relaxed enough to sit and write. But
there are always distractions.

Next to the window of my cabin is a paper birch tree. It had
grown at the edge of the old cellar hole with collapsing stone foun-
dations that I rebuilt to put my cabin on. Each year it hosts aphid
colonies tended by red ants that live in the cabin's roof space. A

highway of ants up and down the white bark attracts a pair of yellow-bellied sapsuckers. Just three meters from me as I sit on the couch writing, these beautiful woodpeckers unobtrusively and silently lap up ants from a trail with traffic both up and down.

Hearing a woodpecker tapping from the cabin wall opposite the birch tree wasn't particularly surprising to me; I assumed one of the sapsuckers was temporarily distracted from the ants and had started sampling wood. Strangely, though, on several occasions when I opened the door to look I saw instead another woodpecker fly off: the northern yellow-shafted flicker. After a time I noticed a suspicious rhythmicity to the tapping. Flickers feed on ants, too, but as far as I knew, they foraged for them on the ground.

A day later, on June 8, I got out of bed at 4:30 a.m. to write up some raven observations from the day before. A little later I heard the same rhythmic *tap, tap, tap, tap, tap* from the same cabin wall. Surely the noise would stop soon. At 6 a.m. it still had not let up, but my patience had, so I gently opened the door to peek around the corner, and again saw a flicker leave. This time I noticed a small hole almost through the wall's outer pine boards: the flicker had apparently been making a nest hole. But the outer and inner cabin walls are separated by a ten-centimeter gap, so if the flickers (I assumed it would be a pair) penetrated the outer board they would meet a bottomless space, with no place to put their eggs.

The next morning at 5:10 a.m. I heard a rustle on the wall, then a light tapping changing to vigorous hammering, which continued nearly unabated for two hours, finally stopping when a second bird arrived. A period of silence was followed by five separate soft, second-long drumrolls, probably some sort of signal. Then total silence. Had the pair gotten through the outer board?

The hole was now nearly large enough for them to slip through, and I was afraid that when they discovered the empty space they would leave and make a nest hole elsewhere. The opportunity to have flickers nesting in my cabin was too good to pass up. I had to do something to help make it happen, and I had to do it without the birds noticing me, in the few minutes they were away.

The potential nest site was too high for me to reach from the outside, but I calculated where it would be on the upstairs bedroom wall. With my chainsaw I removed a section of the inside wall covering its anticipated location. I fixed boards to the bottom and sides below the entrance hole to create a possible nest cavity, cushioned its floor with sawdust and woodchips, and had barely swept the sawdust from the floor, bedding, and clothes and settled in downstairs to wait when the tapping resumed.

By midmorning the flicker was in the house, or to be exact, in its east wall. By afternoon I heard scratchy noises there but no more tapping. The scratchy sounds plus occasional very light and brief tapping continued into the evening and resumed at 5:10 the next morning, and again continued for two hours and stopped when a second bird flew to the wall. As before, I heard the signal drumrolls, which sounded like fingers running over the teeth of a comb. Then all was quiet. I was happy: I knew now that the flickers would stay to nest.

At this point I was not yet living full-time at the cabin, and left it for a few days. Returning on June 16, I was eager to see if I could call it not just the Tree House, as I had until now, but the Bird House. It was the latter! As I walked toward the cabin, a flicker flew out of the hole in the wall. I rushed upstairs, removed the loose panel I had left over the cavity, and looked into the nest. To my joy, on the sawdust and wood chips lay a clutch of seven pearly white eggs.

Flickers normally take about two weeks to excavate their nest cavity, and the male does most of the work. However, the female determines when to lay and how many eggs. Apparently the time until egg laying is not measured from the beginning of nest-hole construction, because this pair had a suitable nest cavity in only three days and egg laying began then. The trigger for the physiological changes of egg production and laying thus appeared to be related to timing of nest-hole availability.

My flickers stopped at seven eggs, a normal clutch size. But that number is below what the species can produce. Flickers are, like chickens, indeterminate egg layers; removing one egg from the nest while leaving at least two in it induces the female (provided she has enough food) to replace the egg. In one case a flicker kept laying until she had produced a total of seventy-one eggs, all the time apparently perceiving she only had about five — not yet a full clutch.

I found the flickers' incubation time peaceful and satisfying. It was a comfort at night to think that a flicker was sitting on her seven eggs about three meters from me, sheltered from the weather. On nights when the pounding of the rain got louder and louder until it became a roar, I felt good that both of us were high and dry.

One early morning after the rain stopped I heard a soft rustling from within the woodpecker hole, then some brittle scratching sounds and several fluttery vibrations. Something was happening. Listening closely with my ear to the wall, I thought at times I was also hearing faint whispering and cheeping-churring voices.

At first light I looked into the nest and saw a heap of tiny naked pink bodies, with empty eggshells scattered among them. The babies made a scratchy-sounding purring noise, except one that made high-pitched peeps.

The sounds they made were the weirdest, most otherworldly I've ever heard, and the most improbable. If I had to describe them I'd say, "Just like you'd expect baby pterodactyls to sound, only cuter." And for that matter, the little pink bodies with their tiny heads on long snakelike necks probably didn't look much different from reptiles either.

In order to watch the birds from up close without their knowing, I covered all the windows to darken the bedroom and inserted

Baby flickers before and after they begin to develop feathers.

a pane of glass into my viewing hole. Later I sometimes removed both the board and the glass and set up my camera in front of the nest to take pictures. At first the parents seemed not to notice that the back wall of their nest was missing, but then they inspected the cavity from top to bottom as though looking for something. They also hung by their feet from the lip of the cavity and looked around my room. I sat frozen in front of them in the dark.

Three days later the chicks had grown noticeably, but they were still surrounded by eggshells. Most parent birds remove shells as soon as, or even before, all the eggs have hatched. This is an evolved anti-predator strategy; even eggs that are well camouflaged with pigments on the outside are white on the inside and therefore conspicuous enough to attract predators if left near the nest. But in a deep nest hole there is scant need for camouflage and thus no rush to destroy the evidence of hatchlings. The eggshells may be ignored, or in time eaten or trampled down.

During the two nights after hatching (July 3 and 4), the babies made upward-inflected whining calls — I suspected they were hunger/begging calls — all night long, reminding me of a flock of bleating goats. The sound was not soothing, and I plugged my ears trying to block it out. I did not know if there was a parent in the nest, and I didn't yet dare to look into it at night to find out, because that would alarm a parent if one was there. The skin over the hatchlings' eyes had not yet parted, but they would have reacted to almost any vibration by stretching their scrawny necks to beg for food. On the night of July 4, when I was fairly sure no parent was with them, I finally did check several times, and there was indeed never any adult present, but they were as noisy as before.

By July 5 their eyes were starting to open. They still bleated occasionally at night, but the sounds were softer and their piping choruses were separated by long silences. After a few more days they began to be quiet all night long. The parents' hygiene behavior at the nest changed too. At first they had picked up and swallowed the young's feces. By the fifth day, the poops were the size of hazelnuts and wrapped in neat membrane-bound packages. The parents no longer ate these fecal packets but picked them up and carried them out of the nest.

The babies' necks seemed grotesquely long and snakelike even before their eyes had opened, and the instant an adult blocked the light at the nest entrance, all their heads shot up and in a chorus of scratchy-churring voices they begged to be fed. The parents delivered food by thrusting their bills into their offspring's gaping mouths and vibrating their heads in a jackhammer-like pumping motion as they regurgitated food deep into their offspring's throats. Each such transfer took about two seconds, and during any one visit the parent might make a dozen or more transfers,

Top: *An adult flicker inside the nest in the cabin wall, looking outside.*
Bottom left: *Flicker delivering food to nearly full-grown young.*
Bottom right: *Flicker reaching between young searching for fecal droplets.*

distributed to several chicks. In the melee of weaving heads I had no way of knowing how much food was being transferred or how many times any one chick was fed.

This "bill-into-mouth" feeding protocol also made it nearly impossible for me to see what kind of food the parents brought to their young. But occasionally I saw spill along the sides of their bills during food transfer, and it seemed to be white ricelike objects. To find out what it was, I followed a standard ornithological method of assessing birds' food. I moved the chicks from the nest to a dark box, loosely wrapped rings of pipe cleaner around the necks of two of them, and replaced them in the nest. Now after the two ringed chicks were fed the food would be detained in their mouths rather than immediately swallowed. Seconds after one of the pipe-cleaner babies was fed I retrieved it and checked the contents of its mouth: a bolus consisting of 140 larvae, 29 pupae, and 47 adult ants.

That sample may or may not have been numerically representative of all payloads these flickers brought back to the nest and fed to one chick, nor were the thirty-two trips the parents made on one day necessarily average. But using these numbers as ballpark estimates indicates that the eventual twenty-two days the babies were nest-bound involved nearly 700 feeding trips, or 100 trips to rear each of the seven young, or about 21,600 ants to fledge one flicker.

The huge numbers of tiny larvae and pupae brought to the nest at any one time should permit the flicker parents to spread the food to several babies in one trip. In most insectivorous birds, however, a meal consists of one large item, and the baby that receives the food is stimulated to release a fecal bolus, which the parent intercepts and removes. But when the food is spread around, the ability to synchronize feeding and defecating makes nest sanitation difficult; if a parent feeds several young in one nest visit, for example, how can it predict from which one to be prepared to intercept poop? I saw part of the solution: the young held it in, and the parent *selected* which one would poop, and when. It did so by gently touching the end of the chick's tail with its bill. Apparently this touching is a signal for the young to defecate, whereas in most birds the voiding response is activated at the head, by contact with the food coming in, so that the parents can attend to the rear immediately after attending to the front. Sometimes the flicker parents also poked around under the young to search for additional feces before hopping back up to the nest entrance.

My close-in observations from the comfort of my chair in front of the nest day after day in early July allowed me to see that when the male came into the nest he typically fed up to five different mouths, checked the bottom of the nest for feces, then left promptly. In contrast, the female usually fed ten or more mouths in any one visit, and she often lingered at the nest entrance for up to twenty minutes, during which time she sometimes reentered, inspected the whole nest, perched next to the young and looked at them closely, hopped back up to the nest hole, perched there a while longer, and then reentered the nest once more. After repeating such a sequence several times she perched just outside the nest hole on the cabin wall before eventually leaving. During one 2.5-hour watch the female spent seventy-three minutes at the nest, the male only eighteen. She did forty-one feedings, he fourteen. She made five trips, he three. For the brood, this averages out to about three feedings per nestling each hour. In another half-hour of watching I counted him making nineteen trips and her only thirteen.

It might appear that the male had worked harder at nest provisioning than his mate, but she averaged three times as many feedings per trip to the nest than he did. Thus, despite his more numerous trips, she delivered seventy-six feedings to his forty-four. She appeared to be selective, first feeding one chick, then deliberately reaching over to feed another, in comparison with the male's apparent random deliveries into any nearby open mouth. During the nest cleaning, in an interesting form of cooperation, he routinely looked for feces on the left side of the nest, she in the center and on the right. Each parent achieved at least some efficiency in their shared caregiving by combining "food in and feces out" with each nest trip.

After a while I got to know one of the chicks. For identification here I'll use male pronouns and call "him" Pipsqueak. He was the smallest and shrillest of the seven, the one that had been still thumb-sized when the largest already filled the palm of my hand. Pipsqueak never seemed to stop calling, at almost exactly once per second, whereas normally the still-naked and -blind young reached up to gape and beg loudly only when a parent blocked the nest entrance.

Pipsqueak's back heaved at each exhalation, with each call. Each time a parent approached the nest hole, Pipsqueak strained the hardest of them all to pop up and reach high on his spindly matchstick legs, wing nubs held over his back looking like arms with elbows held high and hands small and fingerless. He wobbled erratically in straining to stay erect and to hold his head high on his outstretched neck.

By July 12, six days before fledging, the larger of the young were starting to cling to the entrance of the nest cavity to intercept the parents when they came to bring food. At this point the parents stopped entering the nest entirely. The largest of the young then appeared to dominate access to the nest entrance, and hence to get a large share of the incoming food. Pipsqueak was having an increasingly hard time getting fed.

The nest interior looked more and more like a madhouse. The young fought in what resembled life-and-death struggles. Size difference would normally be crucial in combat, but the young flickers had something that mitigated a size disadvantage: a sharp beak. The biggest of them indeed at times occupied the nest entrance, but Pipsqueak too had a sharp beak, and he used it. I saw him jab a bigger sibling at the hole and drive it down, then nab both of the feedings the male parent brought. Sheer motivation was another equalizer. The hungriest were the most willing to stay at the entrance continuously despite having to fight and endure jabs to stay there.

By July 15 the young were feathered, and there was almost always one at the entrance making *kiah* calls to a parent that often perched, seemingly indifferent, directly by the nest hole, or that sometimes vocalized in a nearby tree for minutes at a time. I suspected that the parents were trying to lure their young out of the nest, and that fledging could begin at any time. Wanting to see it, I got up very early for the next few days.

July 17. Until now the adults had appeared at the nest entrance around 5:30 a.m. This morning as usual a chick was at the entrance at that time, and it kept yelling for three hours, but no adult arrived. At 8:30 a.m. a parent finally appeared, and as I watched I suddenly saw *two* flickers fly off. I had in the flash of the moment missed the details, but it seemed as if one of them had just fledged and maybe a parent had flown with it. To be sure, I checked the nest and found only six chicks. So one had indeed left, and it had flown very well on its first attempt — so well that I had almost mistaken it for an adult.

I had a long wait at my post in the cabin before another one left, this time observing the second chick's hind end as it stayed at the entrance for two hours. A parent came to the nest hole, but instead of offering food it just left again. The youngster *kiah*ed even louder than before, then it too flew out and away.

The remaining young were mostly hunkered down almost passively at the bottom of the nest. But on two occasions they suddenly jumped up, only to be viciously hacked at and beaten back down by the new top bird, who was a poster child for pent-up energy. At frequent intervals it stretched first one wing, then the other, then both at once. It leaned far out of the nest hole, pulled back inside, leaned out, and so on. Sometimes it vigorously pecked at its own chest and belly feathers, as if frustrated by not getting fed and yet unable to take the jump to escape the madhouse in the nest below it.

When finally an adult landed at the entrance at 11:29 a.m., another chick jumped up and squeezed into the entrance beside the top bird during the instant when it could not jab to retaliate because its bill was receiving a load of ants. As the parent left, this same larger and ever-vigorous entrance-hog resumed making the *kiah* calls and did so for eight minutes as a parent remained conspicuously in view on a bare branch of a nearby maple tree. The madhouse scramble in the nest hole continued.

There had been only five feedings that day by the time I left my indoor nest-watching post at 1:18 p.m. and went outside to continue a four-hour watch. During that time the female fed the young on five occasions, the male on three. At 6:50 p.m. she made an until-then record of twenty-two food transfers to several of the young. It appeared that the adults had finally relented and ended the enforced fast, perhaps as a way to keep the young in the nest hole for the night. A barred owl came to the woods nearby, perhaps attracted by all the noise. One parent went near the owl and *kiah*ed, then the other joined it in mobbing this predator of easy pickings such as young birds.

In the night I was awakened from a sound sleep by an incessant bedlam in the nest hole. However, from these older birds I heard no vocalizations but only physical tumult. I expected the remaining young to fledge that morning, July 18. The first parent to arrive in the nest vicinity was the male. As usual, he delivered food to two chicks, and also as usual the female made six to seven deliveries during her visit. Both also made trill calls during flight and some *kekkekkek* calls, and the third youngster finally fledged at 9:42 a.m.

This exit from the nest looked similar to a possible previous try an hour before. That time there were vocal exchanges of seven minutes in duration between adults and young. But although one of the young nearly fell out of the hole, it did not launch. The parents seemed to be making deliberate attempts to motivate their offspring to jump. But the youngsters had never before flown a wing beat, so how were they to know if and when they *could* fly? Their unwillingness to jump, and the parents' attempts to overcome their reluctance, were not surprising.

As I watched all day, trying to see every chick fledge in order to learn more about the process, I was unable to differentiate between the two adults except when they landed near the nest, when I could note the presence or absence of the black facial streak that identifies male adults (the young all have it also). The male and female parents visited the nest eleven and ten times, respectively. As always before, the male made only two food transfers per visit, the female five to fifteen. Both were still feeding the young mainly ant pupae and larvae, which they now often spilled because instead of using gravity to help drop the food into the mouths they now had to regurgitate upward to feed the young. A short-tailed shrew was catching the spill on the ground below.

Only three of the seven young had fledged in two days. But surely the other four would follow on this, the third day, July 19. They did, but not without some interesting surprises.

The day's first fledging was at 6:47 a.m. The male parent, which had been perching on a black locust tree at the edge of my clearing, flew to the nest entrance, back to the locust tree, and back to the entrance again, at which point the baby flew out and followed him into the maple woods in the same direction the others had gone.

The next fledging was an hour later. As in the previous fledgings, I heard many *kiah* calls from the woods where the baby and parent had flown. The sixth chick left abruptly at 9:08 a.m., after a long silence.

The departure of the one remaining baby, Pipsqueak, was protracted and tortuous. He appeared at the nest entrance only

FLICKERS IN THE HOUSE

twenty minutes after the sixth had fledged. Unlike the others he was quiet, and he perched well inside the nest hole. No adult came for over an hour. How would they know that a chick was still in the nest, or where their other six young were? Would they remember the last, now silent, one?

Pipsqueak ignored a sapsucker that came repeatedly to pick ants off the column traveling up and down the trunk of the birch tree in front of the nest hole. Like the other young, he also ignored robins and a mourning dove that flew by. He might not have been able to distinguish most birds, but he could tell them apart from flickers.

No flicker came back until 11:10 a.m. And this one, the male parent, didn't go directly to the nest hole. Instead, he perched for six minutes in the black locust tree. Pipsqueak saw him immediately and *kiah*ed the whole time he was there. The parent made a few subtle calls, then left, coming back in ten minutes to trill in flight. Pipsqueak again piped up with *kiah*s, although they were neither loud nor frequent. I waited anxiously to see what would happen next.

Pipsqueak stayed in the nest and got no food until 3:50 p.m., when finally the female landed at the nest entrance and fed him. Both parents returned at 4:26 p.m., when one of them flew to the nest but then instantly left without feeding him. This time Pipsqueak took the plunge: he jumped off, but only fluttered weakly as he planed down and landed in an impenetrable thicket of meadowsweet, fireweed, and tall grass. A parent circled over him once, but no flicker would enter such a tangle, and he fell silent as the parent left. It seemed this might be the end for Pipsqueak, because he would now be unable to see and signal to any parent that came near. To call continuously outside the safety of the nest hole would be to invite hawks and owls, and to stay silent meant to starve. So I went to find him, extracted him from the thicket, and carried him to the cabin. I put him into a wire cage and set the cage under the nest hole where he could see and be seen and fed through the wire. But no parent came, and when I checked on him that night he was lying on the cage floor and seemed unable to move.

At 4:30 a.m. I woke wondering if the parents would come looking for him. He was to my surprise still alive. He had probably been in torpor — a hibernation-like state in which some birds drop their body temperature to save limited energy resources. Suddenly, at 7 a.m., the formerly silent baby erupted with loud *kiah* calls and repeatedly hopped against the cage wall that faced north. I saw nothing in that direction, but a minute later heard the flicker *kek-kekkek* calls. One of the parents arrived, and a vocal exchange ensued, with Pipsqueak beside himself in eagerness. Might he now be able to fly and join the parent? It was worth a try. I tossed him into the air.

Sadly he just fluttered weakly, planing down even more steeply than the day before into the grass. I picked him up and carried him into the woods, where the parent had continued to call. This time when I tossed him he at least landed on the ground where he would be seen, and where he could perhaps gain altitude in a tree by hopping from branch to branch.

At first there was a long silence. But then his *kiah*s resumed, and there was an answer! I left him there. Returning an hour later to check on him, I found no sign of a flicker, only two robins, an adult and a recent fledgling. Sunlight filtered through the green maple foliage onto the brown and moldering dead leaves on the ground. There was silence. Then a hermit thrush sang its slow liquid song, and I returned to the cabin to examine the empty nest.

It contained a layer of pasty, semi-liquid black muck several centimeters thick: decaying feces reeking of ammonia and crawling with maggots. I cleaned it out, hoping for a re-nesting the following spring.

On May 2 the next year, I heard the first flicker of the spring back from migration. I rushed upstairs and pulled off the wood panel that covered the nest cavity. Seeing all in order, I replaced the panel and went downstairs. Several seconds later I heard a tapping that sounded just like a flicker's. I went outside and indeed saw a flicker taking off from the nest entrance. And yet no pair returned to it.

Instead, I soon observed a flicker hammering out a nest hole in a rotting red maple tree about two hundred meters from the cabin. No young were raised there, either — that spring most of the bird nests I examined failed because of continuous cold rains. But the following winter the hole in my cabin contained some chewed remains of red acorns plus one black and several small white feathers. It had probably been temporarily used for food storage by a white-footed mouse and for a sleeping shelter on cold nights by a white-breasted nuthatch. A flicker had stayed around the clearing throughout the summer and into late autumn, feeding along with the robins on the chokecherries as they ripened, and a couple of times I had thought I heard a faint scratching noise at the nest hole at nightfall.

2

A Quintet of Crows

A CORRESPONDENT OF MINE, THE BIRD CARVER LANCE Lichtensteiger, was so enamored of his pet crow that he described it to me as "The most loving animal — I never imagined." He had adopted the crow when it was a baby, after it fell from a nest when a sharp-shinned hawk killed its siblings. At the time Lance already had a gray parrot, which had a vocabulary of over seventy-five words and could sing the first stanza of "Yankee Doodle Dandy." The crow, Lance told me, "learns in minutes, what took years for the parrot." The crow stacked quarters on top of one another and then hid the stack in a crack in the couch. And "when I put my ear to the ground as if to try to hear something, she comes over and does the same — she understands."

In various stages of growing up, I have had three different crows as hand-reared but free companions. I got them as babies from nests that I found in the woods through long observation of the adults and by climbing usually very tall trees to reach the nests. As is necessary and also the chief delight in any bird adoption, I spent hours for and with each crow every day. We became companions. They followed me as they follow one another, and when they were nearby on their solo flights I would sometimes call to a crow in the sky and it would respond by diving down and landing on my shoulder. Some would say they were imprinted on me. I would deny this. It was simply that, given the situation, they had choice, and we were friends. My childhood dream for when I grew up was to live with my crow in a cabin in the woods. If I were to choose now, crows would still be my first choice. I wish every child had the chance to be as privileged as I was.

Perhaps it will happen. There has been a shift in our attitude toward corvid birds. Crows are becoming more suburban, and there are reports of crows leaving objects that seem to be thank-yous to people for feeding them. For example, an eight-year-old girl named Gabi Mann regularly received trinkets such as buttons, jewelry, and bits of colored glass. Her online story inspired readers to post details of their own experiences with crows and to write comments such as "We love our crows," "I fell in love with this beautiful and intelligent creature," and "I treasure the connection."

Unfortunately it is illegal to take a live crow, although it is perfectly all right to take dead ones, after shooting them on sight for target practice. But what is legality, if it is legal to torture a goose or a duck by putting it in a cage where it can't move, shoving a tube down its throat, and force-feeding it to make its liver fatty in order to make foie gras for people to spread on crackers? The Migratory Bird Treaty Act of 1918 applies to crows because some of their populations migrate. But the treaty provides that a species under its auspices may be hunted under regulations preventing detrimental effects on the overall population if there is good cause. Crows are exempted from the act's protection when they "harm livestock" by eating corn. So American crows, *Corvus brachyrhynchos*, are considered great for target shooting. There is no bag limit. There used to be a specific crow-hunting season, beginning in September in some states. But in my state of Maine you can now shoot crows in any number at any time, except on Sundays. Migratory woodpeckers, such as the northern flicker, in contrast, are as far as I know *not* fair game even when they are damaging a home. And I think that is fair and reasonable.

On the night of January 31, 2012, I walked through a foot of new snow. A northeaster was roaring through the woods, but I heard a barred owl nevertheless. At the cabin I built a fire in the woodstove to take the chill off before settling in for the night. The next morning at first light, as always, I heard a raven pair that roost in the pines close by call raucously before leaving on their daily errands. I walked down my hill to drag up the carcass of a stillborn fawn-brown Guernsey calf, brought from a dairy farm. I left it in the sugar maple grove about two hundred meters from the cabin, where it was soon rock-hard frozen and I chopped it open so the birds could feed on it.

An hour later a raven landed in the top of a pine tree, called briefly, and left. The next evening a pair of ravens flew over to the carcass, also cawed briefly, and flew on in silence. The ravens were apparently not interested in the calf, and I later found them feeding on a deer carcass in thick woods nearby. Then a few days later, to my surprise, a group of five common American crows arrived. When I was a boy we never saw a crow in this part of Maine in the winter, and the cawing of the just-arrived crows in April was the first sign of spring. To see them here in midwinter is a treat.

With pleasurable memories of tame crows I had known, I listened to the five crows' calls as they perched in the tiptops of the pines at the edge of my clearing. Within moments they landed on the snow near the calf, and unlike Maine ravens, which usually hesitate sometimes for days before touching a carcass, the gang of five (as I would soon call them) walked up to it and started feeding without delay. In contrast to the squabbling that ravens are prone to, the crows showed no signs of status-posturing of dominance or submission.

By February 14, tracks showed that a coyote and ravens had also come near the carcass. But although both had inspected it from all sides, neither had touched it. They had walked up to it and then veered off. I erased their tracks from the snow to be able to detect new activity in either day or night. A red-tailed hawk also came, flushing the five crows as it landed in a nearby birch tree. The gang of five flew up and dive-bombed it, cawing all the while. After two minutes the hawk left, with the crows chasing it briefly.

The calf carcass was not again of interest to the hawk, but continued to be frequented by both the gang of five and sometimes a pair of crows, though not the two groups at once. Clearly the five were not just random birds that happened to be together at a given moment because of the food. They were a team, and their companionship and solidarity differed so profoundly from what I was used to seeing in ravens that I decided to watch the gang of five more closely to learn about what it is like to be a crow rather than a raven.

One of the questions routinely asked of ornithologists is: "What is the difference between a raven and a crow?" My flip initial answer, which relates to taxonomy, is that ravens are a type of crow. Confusion stems mostly from local nomenclature. All crows (including ravens) are corvids, of the genus *Corvus,* from a common stock originating in Australia. About a dozen species of *Corvus* are distributed, for example, over North America and Europe combined. The one we most often call a crow is *Corvus brachyrhynchos,* the American crow. A similar species over a large part of Europe, *Corvus corone,* is called the carrion or raven crow. The smallest crow is the Eurasian jackdaw, *Corvus monedula,* which weighs about two hundred grams, roughly half as much as the common American crow.

The largest corvid is the common raven, *Corvus corax*, which weighs up to about two thousand grams. Perhaps because of its large size and familiarity to people all over the northern hemisphere, it has garnered a special designation as "the" raven, legendarily known for its intelligence, having a brain capacity roughly double that of the common American crow. Crows can be impressively clever, but only the common raven will fly, at times apparently alone, over the forest emitting a continuous monologue of unendingly varying exclamations and murmurings, then casually tuck in one wing and do a half-roll, as a child might skip while running along. Only ravens will dance high in the sky in the moonlight, or pass snowballs back and forth from one to another in flight (as Andrea Lawrence and Alan Burger have observed), or bring golf balls to their nest.

Another major behavioral difference between ravens and crows is that crows form stable groups of individuals. Depending on context, they often allow their offspring to stay and become helpers at the parents' subsequent nesting. With ravens, in contrast, the young are chased off and/or disperse a few months after fledging.

February 23, 2012. There is still some calf meat left. Sitting at my desk writing, at 2:10 p.m. I hear a crow making short staccato calls. The gang of five has arrived. Four feed while the fifth perches about six meters up in the birch tree near them. It continues cawing, with only short breaks, as those on the ground eat silently. At 2:26 p.m. one leaves the feeding crowd, but it returns in two minutes, so four are again together while a watcher still caws from the perch above them. With only slight variations this sequence, of one watching as the others eat, is repeated three more times until 3:37 p.m. But then, at 3:40 p.m., when three birds are feeding, the three suddenly leave all at once, the other two join them, and the five fly off together down the valley.

The first crow had watched for twenty-two minutes. Others who took over did so for short durations, and they waited for the rest to finish before leaving with them. My impression is in agreement with folklore: crows in cooperative flocks have individuals who voluntarily desist from feeding to keep watch.

The resident pair of crows continued coming to the carcass daily later that month, long after it was nearly depleted and the gang of five had left. Finishing the carcass, the couple then came to peanuts I had spread on the snow by the cabin. On their first visit one of them kept watch and cawed while the other silently grabbed peanuts and then cached them. After a while, presumably when they considered it to be safe, the birds sometimes collected the peanuts side by side.

The crows at no time recruited strangers to share their food — a trait I tested again in mid-March 2014 by putting out a roadkill deer at the same place where I had put the calf two years earlier. A crow pair arrived within an hour, cawed briefly, and then one fed while the other perched silently nearby in a tree. After a while, the feeding crow flew up to perch. The other one started flicking its wings and tail — the signal for arousal or excitement that most reliably distinguishes crows from ravens from a distance — then flew down to feed. After that day the pair came regularly. A single crow also came. It was usually separate from the pair, though it sometimes joined them at the feast. But there was never a crowd; the same birds that had found the food returned to feed on it.

These observations might have seemed mundane to anyone else, but I compared them with those of ravens accumulated over the years at the same place. A deer carcass discovered by one raven was almost always soon shared by dozens of others, and often a cow or moose carcass soon accommodated more than a hundred. Most of the ravens came and went as individuals, rather than in cohesive groups as with crows. Ravens don't recruit others to join in the feast from altruistic motives. Their sharing results from precisely the opposite impulse, a selfish motive *not* to share. Sharing happens because territorial pairs defend their food. For others to get access to the food, they need to recruit a gang and overpower the defenders. Once that goal is accomplished, a free-for-all ensues, with all the birds trying to grab as much as they can as quickly as they can. In this fracas, each raven hauls off one load of food after another and hides it, and when the source is exhausted each tries to steal caches made by others, a task that is not always easy because individuals defend even those bits of food if they can.

In the same period when the single crow, the pair, and the group of five were feeding peaceably one year on a calf and the next on a deer, a quite different scenario unfolded nearby. It occurred in an open snow-covered field by the house of Duane and Nancy Leavitt at the edge of the nearby town of Buckfield, Maine.

The Leavitts had seen a trio of crows around their property for years, and knew them well enough to have named them (Dick, Donald, and George). The three had a group solidarity and were never seen to squabble. But between 7:35 and 7:45 a.m. on February 24, 2014, the Leavitts were alerted by a loud commotion in their backyard. They looked out and watched dumbfounded as two crows attacked a third while several others cawed in the trees near the field.

The three crows on the snow outside the window displayed a flurry of flapping wings and rolling about that the Leavitts described as "a bar-room brawl." After ten to twelve minutes of this, the crows left — except for one that lay bloodied and dead. Ten minutes later a single crow arrived and pecked the dead crow. The carcass was then left *in situ*, and I retrieved it eleven days later when I heard about the incident.

The killed crow was lying on its back and frozen solidly into the crusted snow. Its head, buried in bloodied snow, was partially bare and the remaining feathers were matted with dried blood. Dried blood also marked its abdomen and the base of its right leg and tail. Its chest was dotted with puncture wounds. All traces of tracks from the fight had by then been obscured by a brief thaw followed by a dusting of new snow, but on the new snow there was a set of fresh crow tracks that came within a meter of the carcass.

After thawing the dead crow indoors for a day, I skinned it and was surprised to find bulging breast muscles, not the shrunken emaciated ones I had expected in a weakened bird. This crow had fat on its thighs, at the insertion of its neck, on its belly, and on feather tracts. It showed no signs of disease, prior injury, or starvation. No bones had been broken. The skull showed it to be an adult. It was a male with slightly enlarged gonads. It had sustained massive injury to the head, where skin, skull, and brain showed red from hemorrhage. Both eyes had been punctured. But not a speck of flesh had been removed from anywhere on the body.

The crow had succumbed to multiple pecks to the head in a deliberate and sustained attack. Numerous peck marks in one small area of the left breast muscles showed no evidence of bleeding; they had probably been inflicted when the victim was unable to move or after it was dead.

To my knowledge there is only one extended discussion in the literature of the rare and mostly unsubstantiated anecdotes suggesting that groups of crows gang up to attack individual crows. A commonly supposed reason for this behavior is to drive away or kill a weak or injured bird for "status enhancement." I'm skeptical of this rationalization, because although the risk per individual participant is lower in a gang attack, the potential enhancement of status, if it were to accrue by killing, would be small, especially if it was unclear who among the crowd had done the damage.

However, there were some clues to the killing that bracketed the possibilities. First, the attack occurred near a feeder and not during a time of starvation; the crow was not killed for food. Second, the killed bird did not appear to have been weakend: it was of average to large size for a northern American crow in winter, and the autopsy indicated that it had been in excellent physical condition. These facts suggest that it may have been the aggressor, one able and willing to risk a fight, rather than a victim of a gang attack waged by individuals vying to score a kill to enhance their status.

The victim was a male strong enough to risk a fight, and he

would have done so only when victory could have significant long-term benefits, such as reproductive ones. Had he tried to secure a mate by ingratiating himself with the trio of crows that had long resided in the area? The breeding male of the trio would have had more to lose than a challenger, because his mate was real, not potential. A scenario of two males' cooperating against a would-be interloper presents an alternative hypothesis to the prevailing one of crows' attacking a weakened individual simply because it is weak.

We humans find crows to be loveable because they have the mental capacity to bond (usually with kin). We notice this quality when we have one as a pet, to which we become a surrogate not only of the crow's own kind but also of its family.

Social units evolve under selective pressure to garner help to gain an advantage in competition against other groups. Usually the stronger the ties to kin, the greater the intolerance of nonrelatives. So by its very nature, being loveable in bonding to members of the group implies discrimination that can lead to aggression toward others. The discrimination that results from favoring individuals for cooperation can scarcely exist without the discrimination against others, any more than light can exist without shadow.

3

Getting to Know a Starling

I HAD GONE FOR A SWIM IN OUR FAMILY'S FAVORITE POOL BY a big rock in a bend of the Huntington River in Vermont. It was a hot day in August, and after a refreshing dip in the clear swirling current, as I strolled homeward along the hemlock-shaded path by the river bank, I felt a bird flutter over my head. Startled, I looked up and saw a lone starling perched in a honeysuckle bush next to the path. After a few more steps, I stopped short, thinking there had been something deliberate in this bird's coming close to me. People routinely walked along the path to and from the swimming hole, but I had not heard of anyone else being accosted by a bird.

On impulse I picked a few berries from a nearby bush and held them out to the starling. It fidgeted. I talked to it, leaned forward, and talked some more, upon which it fluttered onto my outstretched hand, grabbed a berry, and flew back to the bush. I picked more berries, and we repeated the give-and-take maneuver.

Starlings routinely travel along the ground in flocks, turning over leaves when foraging for insects. Testing the bird further, I knelt and started scraping leaves from the ground, trying to mimic starling behavior. Amazingly, it flew down, landed beside me, and acted as though it was looking for food where I had scraped off the leaves. I had never before achieved such immediate empathy with a wild bird. I had to get to know this one better! I decided to try to catch it and take it home.

When I again held out berries, the starling landed on my hand. But as I slowly drew my other hand closer, "he" flew off (I was unable to determine the bird's gender but will refer to it as a male). On his next landing I closed my fingers, trying to grab him by his legs, and managed to snag him by a toe. He screamed loudly until I cupped him securely in my hands. And I held him that way as I jogged the three kilometers home.

Put into an old parakeet cage after his ordeal in my hand, he seemed relieved: he shook himself and started to preen. He showed no fear of anyone in the family, and got himself adopted into a home where he would be housed and fed through the coming winter. "Pretty slick," I thought. And so "Slick," token of both his appearance and his behavior, became his name.

Slick, a common starling, *Sturnus vulgaris,* lived in his cage next to a window in our kitchen/living room. First thing in the morning when I turned on the light, he faced me and preened nonstop for about ten minutes. He reached for feathers from his back and pulled them through his bill, then did likewise for those on his breast, flanks, and throat. He angled one wing to the side and pulled shoulder feathers through the bill. He'd fluff himself, shake, go sleek, then puff out the feathers on his head and shake that. He'd stand on his right leg and stretch his left leg and wing, then stretch both wings over his head simultaneously, then raise his left foot over his left wing and vigorously scratch the back of his head with a toenail. After repeating these steps over and over, he'd lift his left foot to scratch his chin while closing his eyes, opening his bill, and uttering a tiny squeak. He might sneeze once or twice and wipe his bill on his perch. Then he'd lower his head down into his shoulders and look around.

Slick's feather care seemed to be done for its own sake, for fun. And nowhere was the drive to have fun more evident than in his bathing routine. Bathing was his passion. Seeing water run from the tap in the kitchen sink sent him into a frenzy to be let out of the cage. I obliged him at least every three days and sometimes up to twice a day. I let the water run into a soup bowl. Whether the bowl was dirty or clean, he jumped in under the running water, squeaked as if in ecstasy, bent his legs to get deeper, and whirred his wings like an egg beater in high gear. The spray flew several meters, soaking the floor and counters near the sink. Then he jumped out dripping wet and fluttered back into the cage to shake vigorously and preen until dry.

His grooming regimen seemed extravagant, but his appearance was worth the effort. From a distance he looked black, but from up close he glistened in sheens of metallic green, purple, and blue. The feathers on a starling's breast, head, and neck are purple, and those on its back are green. Slick's belly was blue with green, and just below the purple of his neck the feathers were tipped with light yellow, as if individually dipped in cream. As is typical of starlings in winter, his back feathers were fringed with light brown, so that he looked like a metallic rainbow with dots of cream and brown.

Starlings wear different garb in different seasons. Juveniles in their first summer are an undistinguished gray-brown color and their feet and legs are a dirty brown. When molting into adult plumage in late winter/early spring, they have whitish spots at the tips of their feathers, and by spring they lose the spots as the feather tips wear off. Their brilliant feather sheens then show, their bills become bright yellow, and their feet and legs turn orange.

After his morning grooming, when Slick saw me begin to

prepare breakfast, he hopped around excitedly in his cage while emitting raspy, upward-inflected squeaks. I generally went to the cage and handed him a few tidbits to snack on, to get him to shut up.

A starling's most puzzling peculiarities show up during feeding. Most birds use their bills as picks or pincers. But Slick would stick his bill into my ear and then spread his mandibles apart rather than closing them. This odd bill movement is suited to flipping dried cow-patties to find dung beetles. It is also suited to lifting fallen leaves to check for worms underneath, opening crevices, and spreading grass apart to expose insects and other food. But it was misplaced when he was poking around in my ear.

A second oddity Slick shared with other starlings was an inability to hold food in place with his feet in order to manipulate it for feeding. He was a physical contortionist when it came to preening and scratching himself, but he could not catch on to the idea of clasping a piece of food with his toes. Blue jays and chickadees routinely use their feet to secure a nut or sunflower seed against their perch while hammering it open, or to grasp a cracker while eating it bit by bit. A nuthatch steadies food by inserting it into a crack in a tree's bark and using the crack as a vise. But Slick, when given a piece of food such as a cracker, was at a loss as to what to do. He squawked in frustration, flew back and forth from perch to perch, flung the food item around wildly, shook his head, and sometimes battered the cracker against his perch. But never once did he step on a piece of food to steady it and then try to pry a piece off.

Slick was an opportunistic omnivore. Grapes, raw cabbage, lettuce, scrambled eggs, raw diced carrot, fresh bread, cooked potato, blueberries, yogurt, his own feces, raspberry jam, raw and cooked hamburger, Cheerios, rice, peanut butter, apple, worms, grubs, and insects were all on his bill of fare. But his begging for food was most insistent when he saw me with a cluster fly in my hand. There was no shortage of these flies moving sluggishly in our house in the winter, along with the so-called variable ladybird beetles.

Ladybird beetles are brightly marked in red, orange, and black, and when handled roughly they exude a noxious smell. They evolved the smell, and the bright colors that advertise their noxiousness, to discourage birds and other predators from eating them. I knew this but offered Slick a ladybird beetle anyway. To my surprise he took it, although he had eaten his full dish of food scraps that morning. He shook it, then ate it with little hesitation. I offered another, and another, and eventually twenty-two ladybird beetles in a row before he began to reject them. He beat most of them against his perch as if to shake off their chemical secretions, but in the end he ate them anyway.

The next day he again ate all his food scraps, yet he also consumed eighteen freshly captured ladybird beetles. He gulped the first few quickly, but before finishing them all, he occasionally gave his head a rapid shake after swallowing. Later-presented beetles he beat against his perch before ingesting them, and finally after the eighteenth he started dropping them, seemingly inadvertently. Wondering if he was satiated, I brought in another handful of beetles along with some fresh cluster flies. Now when given a beetle he either ignored it or wiped it against his perch and then flung it aside. But of the forty-five cluster flies I had picked off an upstairs window, he ate every one, right after the other, never once wiping one against the perch or dropping it.

Here was a welcome potential solution to the cluster fly problem, except for the inevitable aftereffects in a free-ranging bird. After Slick's mega-feeding I let him out of his cage to fly around in the house, and he quickly landed on or in my wife's ample hair. This excited them both and caused the starling to produce an unwelcome deposit from his cloaca.

. . .

A flock of starlings.

Starlings are not well liked in America. In the non-breeding season they may roost in flocks of hundreds of thousands, each one singing engagingly but all together making what to us is noise, all the while defecating on everything below them. They are an all too successful invasive species that often usurps bird boxes meant for bluebirds. From two hundred of them released in Central Park in the 1890s, they now number over two hundred million and range across the whole continent. They accomplished their population expansion a lot quicker than did their human counterparts from Europe.

Even when in their huge winter crowds, starlings sing almost constantly and remain vocal long into the night. What is irritating and obnoxious when practiced by the many, however, can be a delight when performed singly. Birds' singing has important practical functions, but because starlings practice it so easily and often it is most likely also fun for them.

Starlings adapt their vocalization to what they hear, and have been known from antiquity for their vocal talents. People who have had them as house pets report that it took their hand-reared birds from a few days to months to incorporate portions of tunes, ranging from Mozart concertos to "The Star-Spangled Banner," into their vocal repertoires. They also mimicked words and short sentences, including "Give me a kiss," "I'll see you soon," and even "Does Hammacher Schlemmer have a toll-free number?"

Slick did not know enough to grasp a cracker in order to eat it. But recent laboratory studies are thought to show that European starlings can "accurately recognize acoustic patterns defined by recursive, self-embedded, context-free grammar" and "reliably exclude a-grammatical patterns." In other words, they can grasp grammar. Pretty slick.

I made no attempt to teach Slick how or what to sing, but all winter he sang almost continuously from mid-afternoon through the evening and long into the night, stopping only when I turned off the light. These untutored monologues were punctuated only by occasional pauses of a few seconds. They were a seemingly endless string of the usual starling trills of various pitches, tinkles, whistles, twitters, growls, and squawks. Thrown in also were phrases of a common yellowthroat warbler song, a wolf whistle, and the ring of a telephone. These sounds were his private repertoire, and they may have reflected some of his past. My occasional whistling at him might have accounted for the wolf whistle. The warbler song meant that he had been near open, swampy, or disturbed habitat in spring. His lack of human words indicated that he had not grown up in a situation where he bonded with people. As the biologists Meredith J. West, A. N. Stroud, and Andrew P. King wrote in an article on starlings' vocal mimicry, starlings listen actively to and mimic only those humans with whom they interact socially in shared companionship. So I suspect that even if

Slick had had human keepers before me they had not interacted with him much.

I recently saw a YouTube video of a starling named Beakie with an impressive and varied repertoire of songs, words, and whistles it had learned over ten years from its keeper, who often talked and whistled to it. The most famous starling in human company was the one kept by Wolfgang Amadeus Mozart in Vienna. Mozart reputedly was walking on a street when he heard a caged starling sing a close rendition of his own Piano Concerto No. 17. He bought the bird on May 27, 1784, and made a record of the purchase that said "Vogel Stahrl — 34 K — Das war schön!" (Starling Bird — 34 Kroner — That was wonderful!) When the starling died, three years later on June 4, 1787, Mozart buried his pet in his backyard and wrote a commemorative poem about it.

A starling singing and gesticulating with his wings.

Slick craved company and was not deterred in the least by the hustle and bustle of even the loudest party, staying cheery in the middle of it. Such tolerance was probably related to starlings' social evolution, in the context of being with sometimes hundreds of thousands of jostling, noisy flock mates, especially in the balletic synchronous flight maneuvers they achieve when returning to their communal roosts in the evening. A starling needs not just a mate and/or a companion. It needs a crowd. To be a starling is to perform airborne dances with myriad others, tracing elaborate syncopated flight patterns in the sky. We call these gatherings "murmurations."

When Slick felt threatened I could almost literally not get him out of my hair, although when relaxed he became a pleasant companion on my shoulder. I was reluctant to let him loose in the winter, when there was no nearby flock for him to join. But in early spring, for a trial, I walked into the yard with him on my shoulder. He stayed perched on me the whole time. I tried it again a day later. This time he flew off and landed in the cherry tree next to the porch. I stayed on the porch and waited to see what he would do next. After a few minutes he returned to my shoulder, then flew through the open door into the house.

When the weather improved later in the spring I took him outside more often. One day he disappeared, and we never saw him again. I still miss his cheerful song, but I could not and would not have kept such a social bird cooped up in a cage unless I could have guaranteed him at least an hour of daily attention to bring out the true starling in him.

Some animals, especially social birds like crows and starlings, require constant companionship, sometimes for years on end. This commitment is not for most people, but wild pets can be an inspiration to others.

4

Woodpecker with a Drum

WE ASSUME THE REASON MALE WOODPECKERS DRUM ON wood is no great secret: they do it to attract females. But although I've seen and heard flickers and hairy, downy, and pileated wood-peckers do their instrumental renditions of birdsong as long as I can remember, I don't recall ever seeing females attracted, or males' territories expanded, as a result.

The yellow-bellied sapsucker is hardly an exception when it comes to drumming behavior. It is, however, one of the few wood-peckers that do not excavate dead trees for grubs. Instead, it taps live trees to lick sap and to feed on insects that feed on the same sap. Unlike any of the other half dozen species of woodpeckers in Maine, it is so intent on drumming loudly in the spring that it opportunistically uses not just dry branches but also tin roofs and metal stovepipes for sound amplification. Aside from that, this species, with its striking deep crimson throat and top of head, pale yellow underside, and bold black and white markings, is knockout beautiful. For all of these reasons, I was pleased to find an excuse to get to know it better.

That excuse came to me unexpectedly around 7 a.m. on my birthday, April 19, in 2012. In the clearing by my cabin is a yellow delicious apple tree at least a hundred years old. Sap still runs in its living wood, but age has hollowed its trunk and thinned parts of its dry-wood walls, creating an excellent drumming place for a woodpecker. And this tree has, thanks to a bear and me, an added attraction for a sapsucker.

The bear had climbed the tree the previous summer to reach early-ripening yellow apples. It had damaged some of the brittle branches, and to protect the tree against a second raid I tacked steel sheathing around the hollow trunk. A sapsucker just returned from its winter home discovered that the metal could serve as an amplifier to make a super drum.

Before that day, sapsuckers had rarely landed on the apple tree, which stood about fifty meters into the clearing and a hundred from the cabin. They generally stayed in the woods, only occasionally venturing to the edge of the clearing and drumming on the cabin's stovepipe. On this gorgeous spring morning I was sitting next to the wood stove, enjoying my first cup of morning coffee and writing up some notes, when I was jolted by the most resounding sapsucker drumming I had ever heard. From my window I could see the male bird (it has a red throat bib, the female a white one) on the apple tree.

One *rat-tattat-tatatattat-TAT* drumroll followed another in rapid succession as the sound went on, and on, and on. After some drum sequences two additional sapsuckers landed on the same small tree. The drumming then stopped, and I heard screechy calls and saw a lot of posturing by the birds, then the three flew off into the woods. Almost immediately one came back to the same spot, only to leave in a few seconds. Why did the others come and all three quickly leave? I had never before heard such a woodpecker din, nor seen other woodpeckers come to it. To have the ones that had been attracted then immediately leave seemed puzzling.

Minutes after the sapsucker started that morning I grabbed pencil and paper and went outside to record what else might hap-

The old apple tree with metal flashing that served as a "super drum" for the sapsucker.

pen. I realized I had caught the moment when a sapsucker found the mother of all sapsucker drums, and had an opportunity for making a discovery. I had found a question that would keep me busy for a long time. Why does a woodpecker drum? I had assumed that the drumming, which I had also heard in the fall, winter, and spring, was a territorial advertisement, like an aural No Trespassing sign. But here, instead, others were *attracted* to it. I needed to take detailed notes to create a foundation with which to compare later events, because I had no idea what might be relevant.

At 7:51 a.m., after a short break, the sapsucker returned to the tree and produced one drumroll after another for five minutes. Then he left. Eight minutes later he came back and drummed briefly, stopping when another sapsucker arrived — a female. He left in moments, and she then also left. He returned at 8:39 a.m., and after ten drumrolls a female again arrived, and again neither stayed more than a few seconds.

At the end of five hours he had completed seventeen separate drumming sessions, for a total of seventy-six minutes. His performances had dramatic effects. Nine times other sapsuckers came: five times lone females; once two females; once a male; and twice three birds at once, which left before I could identify their sex. The females, just before arriving, seemed to announce their presence in the nearby woods with a relatively soft drumroll and/or a vocalization. The one time another male came, the two males appeared to be totally indifferent to each other, and both, when they left, flew in the swooping pattern typical of woodpeckers.

His percussion solos continued for up to forty drumrolls, one after another, before he left for a half-hour and then came back. Curiously, whenever a female arrived he would instantly stop drumming and they would have a brief face-off with vocalizations and displays of puffed-out head feathers. He would then leave, not in the normal swooping flight of woodpeckers, but rather in a fluttering, almost mothlike trajectory. The female would hesitate a moment or two, then fly into the woods in the same direction he had taken.

It didn't make sense. Why did he leave just when he had her attention?

I was hooked. Something interesting was happening, and it had become interesting fast. Clearly the drumming couldn't be directly related to mating: these birds had just finished their spring migration and would not mate for a month or more, after they had hammered out nest cavities, and their eggs were available for fertilization.

I was up in the dark at 5:05 a.m. the next day. A normally nocturnal male woodcock was still doing his sky dance and *peent*ed after landing in the field on the bare spot next to the apple tree. Robins already sang. Chickadees *dee-dah*ed. The phoebe, winter wren, hermit thrush, and blue-headed vireos were singing. No sapsuckers seemed to be stirring.

At 5:40 a.m. a male finally arrived and started to drum. The sound carried for a kilometer or more on still days, and soon there were other drummers in the woods, but his performance was the most vigorous. Responses seemed to come from two directions. One was of ten drumrolls, the second of only one. But he continued on and on — to a record of forty-seven consecutive drumrolls. Then he left.

As on the day before, a female came twice, and each time he left in fluttering flight. During his flight into the nearby woods at 8:26 a.m., I heard screeching and followed him to what looked like a probable nest tree: a poplar with the signature hoof fungus that sapsuckers preferentially choose to excavate for a nest site. However, in a short while he returned to his drum at the apple tree.

I thought I was beginning to see a convincing pattern. But why, I wondered, did the male come back repeatedly to his drum, then leave almost the moment a female arrived? And why, when he left, did he fly not in the normal woodpecker style but rather in a showy fluttering flight? The female at times followed him, not the other way around as I had expected. None of my evidence would make sense unless I could understand what was at stake.

It was clear from the first day that the drumming did attract females, and also that the male made an effort to drum loudly. I had watched as he tapped all around the dead stub of the old apple tree, as though searching for precisely the right spot to produce the most dramatic effects. Then he came back again and again and hammered at that same spot. If he was trying to produce high-quality drumming, and if it served to attract females, then he must be trying to provide something the females wanted or required and were looking for. It could not be just loud noise, as such, because that could do the females no good. His leaving every time a female came made it clear that, at least at this time, what the females were looking for was unlikely to be mating. I wondered where he was going every time he left followed by a female. In my one pursuit of them through the woods I saw no mating.

It was perhaps time for *finding* mates, but still far from mating time, which is near egg laying, as mentioned previously. Long before that these birds needed to hammer out nest cavities. Trying to think like a female woodpecker, I considered qualities I would look for in a mate. Why would I choose a male who made a lot of noise by pounding on a piece of tin on a half-dead apple tree?

The drumming, as such, seemed like a useless activity. It therefore had to convey a message. I thought of the chickadees that were now traveling in pairs. The males make soft peeping calls, and now and then offer their mate a nuptial gift by feeding her an insect. The female then begs, fluttering her wings like a baby bird and mimicking a baby's begging. By her actions she is testing the capacity of a potential mate to feed her young. If he responds to her baby signals by offering her a caterpillar, the male shows her that he is capable of feeding them. A chickadee pair may have to raise as many as eight babies at the same time, and she needs help for the monumental tasks of egg production and the feeding of fast-growing young. It occurred to me that the most difficult

A sapsucker in early spring making sap licks in a quaking aspen crown, and a chickadee taking sap.

and time-consuming task a *sapsucker* female faces may not be the feeding of the young, but the making of a nest hole.

Sapsuckers do not chisel into wood for food like most other woodpeckers. Instead, they nick the bark of live trees to make the sugary sap run, and they lick it for fuel and also eat the insects attracted by the same food source. Excavating a nesting cavity out of the solid wood to create a safe home for their young is thus, for them, a huge job. It is one not unrelated to the feeding of nestlings. Birds in open, exposed nests must raise their young quickly, because as long as the young are in the nest they are vulnerable to predation. That requires fast growth, which in turn requires the parents to deliver massive amounts of food to the babies in a short time. But when the young are in a safe place such as a solid wood cavity, the food-gathering pressure is greatly eased, because the amount of food open-nest young need within a week or so can be supplied over two or three weeks instead. Thus a female sapsucker needs a mate that is strong, able, and willing to hammer in wood, to help her make a nest cavity. And what better indicator of these qualities is there than the volume and duration of his drumming?

I was excited; I had a hypothesis. If it was accurate, my hypothesis would indicate that the sapsucker *males* were the primary nestmakers. This would be very unusual, since almost exclusively female birds make the nest while the males are passive onlookers. (There are exceptions, though, such as weaverbirds and wrens, where the male starts the nest, which serves as a sexual attractant, and the female finishes it and thereby accepts him.)

Predictions can bias and mislead, if they come too early. But once you have a data set, there is nothing like having a prediction for making progress in at least a tentative direction. If the facts match, you are a step closer to solving the riddle. I now had a hypothesis, and if it was a correct one, it had to match not just one but numerous facts of the natural history of these woodpeckers.

In Vermont I once made, for the fun of it, a survey of 176 live poplar trees along the woodland roads I ran on. I found twelve with the *Fomes* hoof fungus, of which five had sapsucker holes, while there were no sapsucker holes in the remaining 164 trees. I concluded that most sapsucker nest holes are in live aspen trees that have a fungus-softened core while the rest of the tree is alive and solid. The *Fomes* hoof fungus, which infects a tree's center and leaves a hoof-shaped fruiting body on its trunk, serves as a marker of a tree where a nest hole can be excavated. If a sapsucker chooses just any poplar tree for a nest site it may face a very difficult if not an impossible task, but if it chooses a tree with a fungal body on it the task will be much easier. However, the fruiting bodies of this fungus are not obvious to an observer. Thus a sapsucker male that finds a suitable tree and shows it to a potential mate is providing a service to her even if he doesn't make the nest hole. I was eager to find out if sapsucker males do lead females to suitable nest trees, and also if males take the lead in excavating the nest hole. It was now the beginning of the sapsuckers' nesting season, and I needed to observe them not only at the drumming tree, but also in the surrounding forest.

The leaf buds had not yet unfurled, so it was still easy to see between the trees and not difficult to find and follow the birds. It also helped that they were noisy at this time; the sapsucker pairs seemed to keep in contact with each other by calls and drumming. I had not known that females also drummed. Compared with the males' resounding hammering, theirs was more like a tapping and had a different sound pattern, although the males' varied as well.

I started at a sap lick on a white birch tree. A female was on the birch, a male on the next tree over. They both made a couple of drumrolls, and I presumed they were a pair. I then headed toward the clearing and the apple tree to look for others. I soon found them. Hearing a commotion, I ran to it and saw three sapsuckers receding into the distance. My crashing through the woods ended up as a nearly hour-long woodpecker chase.

The trio would fly through the trees and stop somewhere for animated interactions, and usually by the time I caught up to them they were off to another location, where their screeching and drumming would start all over again. I found myself going up and down and sideways across the east side of a hill, but when at times I did catch up I could not hold my binoculars steady enough, nor would the birds hold still enough, for me to reliably differentiate the female from the male. But after a while I realized that I was covering and re-covering the same ground, returning several times to the same patch of poplars in the woods just below the apple tree. I had heard no drumming from there, so I suspected that the three birds I was chasing were the drummer and the females that had been coming to him.

After another wild chase I remained at the poplars to wait for a possible interception there. Soon a lone female did arrive, but I lost sight of her. Then I heard a brief drumming here, an answer there, and also screeching vocalizations. I was determined to find out who was chasing whom, and after a while I found that in three chases it was males apparently trying to chase off another male; the entourage I was following included two males and at least one female. And then I found the epicenter of it all: a poplar tree with three sapsucker nest holes. This was the one tree a male kept returning to repeatedly.

Sapsuckers often return year after year to the same tree, where they annually excavate a new nest hole. The nest holes on this tree were from previous years, and there was, so far, no new one. The female drummed near this tree, and the male was at one point on the same limb with her. When just the two were there, there were no chases. Therefore, it was males chasing males, and females accompanying their males.

One male looked into one of the old holes of the aspen tree that seemed to be the sapsuckers' focal point. Additionally, he tapped here and there on the tree trunk, possibly testing it. Eventually I noticed a yellowish patch on the tree trunk a half-meter beneath an old nest hole. I didn't understand what it was until I saw him pecking there for several minutes and realized he was removing the gray-greenish outer bark to reveal the light-colored inner bark, in a pattern roughly the shape and size of a nest-hole entrance. The tree was thus marked, as though with initials, as his. He had staked out a nest tree.

While he was busy at the tree, another sapsucker at least a hundred meters away made screeching calls, and the nesting male — I later identified him as the drummer — instantly launched himself in typical woodpecker flight toward the sound, and I heard another vocal interaction as he tried to chase away the other male.

After pursuing the birds through the woods like a wild man, I had at last found their focal spot, which would soon turn out to be their actual nest tree. I had to leave to go back to Vermont, but when I returned nine days later the male was working deep inside the tree at precisely the spot where the outer bark had been removed. A second question occurred to me: Does the male or the female do the heavy work of making the nest?

During my nine days in Vermont I watched another pair of sapsuckers. I found four different trees where nests had been started, as well as one that had fourteen old (or partial) nest holes from previous years. I watched as a pair of sapsuckers examined them, behavior which suggested to me that they were interested in nesting. Indeed, they began to work up to ten minutes at a time to excavate what looked like a probe hole in a very thick bigtooth aspen with the hoof fungus. The pair worked there briefly, but then abandoned that tree and moved to a nearby quaking aspen (also with the fungus). The male started working and soon had a hole deep enough that his entire head disappeared into it. They abandoned this tree as well, most likely because, as I discovered, it was hollow — the fungus had rotted out the interior and there was no "floor" for a nest. Clearly, finding the right tree can be an ongoing process, not just a one-time event. And the female is far from passive in the process of choosing and making a nest hole.

What actually happens, as I learned by chasing sapsuckers around the woods for a week, was even more interesting than I had anticipated. I saw a female fly directly to the nest hole a male had begun, perch in front of it, and make several *hey* calls. She ducked her head into the hole again and again, at least twenty times in succession. Then she ducked in deeper — up to her shoulders. I saw her tail quiver and her body vibrate as if she were pecking to excavate. She continued for eleven minutes, pulled out of the hole, made four loud screeches, and flew off. In three minutes she was back at the hole and perched there silently. The male arrived, flew to her, screeched, and left with her following. Next he returned to the hole and without the slightest hesitation slipped in and got to work. He reached down far deeper than she had, so she must have merely tapped near the entrance. She then visited him, again perching silently nearby. He hammered loudly without pause for twenty-five minutes. She left after a while, came back, screeched three times, and resumed her perch at the hole. He came out and, as before, they both left in the direction of their sap licks in a sugar maple tree. Soon he reappeared and continued his excavation work for nineteen minutes before leaving to refuel at the licks. He returned at noon and worked for forty-one minutes straight. I made four more spot checks on the nest that day, until 7:45 p.m., and watched two more of his work sessions, lasting twenty-four and nineteen minutes.

The female was a frequent visitor at the nest site, but mostly as an observer or inspector, and when participating in making the hole she did so feebly. My prediction was confirmed: the male does most of the work. I never saw him feed her except indirectly by creating the sap licks. The female, though, also works indirectly for their reproductive effort: while he is excavating, she is collecting insects, gathering the protein she needs to produce their clutch of eggs.

As I later found by observing a fourth attempt to make a nest hole in another quaking aspen with the fungus, the female shows interest in if, where, and maybe by whom a nest hole is being made, and indicates her approval by token participation, which seems to signal the male to take over and do most of the work. When she came to inspect, he would resume hammering as though with renewed vigor. I conclude that the male's drumming attracts the female, and that when she arrives he leads her to his previously found nest site. If she approves, she lets him know by offering token help, and he then begins excavating in earnest.

I eventually saw mating. The female was perched directly in front of the nest hole. Expecting her to go inside, I waited, and waited — as she just sat there, preening her feathers. Finally, after three-quarters of an hour, she started to tap weakly all around the outside of the hole. She peeked into it and tapped some more, again very lightly. The male then arrived and flew to her, and she began to hammer loudly as though using the nest hole as a drum. He fluttered away in his display ("follow me") flight to a nearby branch. He came back, she again drummed, and he again fluttered off. On his third try she followed him to the branch, and they copulated there. Then she left immediately and he entered the nest hole. The tips of his wings showed at the entrance as he worked, indicating that the nest was perhaps half finished. She was presumably busy as well, but in the forest. Preparing to lay five to eight eggs required a great deal of protein; she had to hunt for insects. He, meanwhile, fueled up regularly at sap licks he had made. He always had several close and handy.

When after about sixteen days the nest hole was almost completed, the female helped by throwing wood chips out of the entrance. As I had expected, the male's drumming had by then almost stopped. It was time for the beginning of incubation, when the sugar maples and beeches had leafed out, restricting visibility through the woods. The pair took turns incubating the eggs. And they played what looked like equal parts in feeding the noisy young.

Like all woodpecker babies, the hatchlings made a racket right from the start. Young woodpeckers can afford to beg loudly and almost continuously because their nest, unlike those of most birds, is a fortress, one constructed at great cost and with expert ingenuity by both parents. The babies of other woodland birds, in flimsy, quickly made nests, are vulnerable and must stay silent. They beg only briefly and weakly, and only when a parent is directly at the nest.

My observations now fitted into a consistent story, one that still gives me pleasure whenever I hear a sapsucker drumming in the early spring. Sapsucker males are amazing drummers who go out of their way to be loud. Particularly in sapsuckers, which are not adapted as routine wood excavators, the female may need help to make the nest hole, and may look for evidence that a potential mate is both willing and able to help with this crucial step in the reproductive process. Drumming vigor, as determined by its loudness, would be an excellent measure linked to mate choice by female sapsuckers. But only if the drummer actually "delivers" by doing the work of nest-hole building.

To find out what others had discovered that might refute or confirm my hypothesis, I went to the obvious two sources: Arthur Cleveland Bent's *Life Histories of North American Woodpeckers* and Lawrence Kilham's *Life History Studies of Woodpeckers of Eastern North America*. In both accounts I found confirming observations as well as conflicting interpretations. To my relief, neither work referenced anything like my hypothesis about drumming, even while both largely supported what I had seen. There was confirmation that males helped build the nest holes, but nothing regarding why or how this might be related to drumming or why male sapsucker nest-hole-making might be different from that of other woodpeckers.

The observations in 2012 led to others the following spring, but these were in more detail and with an experiment added. In large part they confirmed what I had found the previous year, so here I summarize only the experiment and its outcome.

The nest of the drummer and his mate in 2012 had been approximately two hundred meters from his drum on the old apple tree. The parents and fledged young later came to the birch tree by my cabin window to pick off red ants, which had a trail going up the trunk to the aphids they tended on leaves in the tree's crown. Thus the sapsuckers were getting a protein supplement along with the sugar the ants were carrying back to their nest (which they got from the aphids, which got it from the tree sap). This was sapsucker heaven — with drum and food source next to each other — and I wondered what would happen if the nest tree, too, were located directly at the male's drumming site. Would he still fly off into the woods whenever a female arrived? Or would the drum's proximity to the ideal nest site keep him there? If he did not leave, would his staying induce the female to accept the location as a nest site?

Expecting the sapsucker pair to return in 2013 to raise another brood of young, I felled the poplar in which they had nested and (with a little help from friends) set up a four-meter length of it, which included their recently used nest cavity, at the super drum at the apple tree. When a male sapsucker arrived the next spring he immediately went to the drum. He also immediately took an interest in both the poplar stem next to him and the hole itself, which had likely held his nest the year before. A female quickly joined him, examined the poplar trunk, and peeked into the hole.

Sapsuckers did not, to my knowledge, reuse old holes. As expected, this male started to make a nesting cavity in the nearby woods. With no poplar close by, he chose to make a test hole in a dead maple tree, but when the female seemed uninterested he quit and started drumming at his drum. He then returned to the used nest site and began tapping inside it and throwing out small chips and debris. And to my surprise, the sapsuckers did nest in that old hole.

Watching them there gave me great pleasure — until, one hot summer day, I saw both of them picking at the tree in a bizarre frenzy and realized that they were picking at ants. They were not just eating the ants but also regurgitating them and flinging some aside.

A colony of the red ants had found the nest just as the young were hatching, and before I noticed what was happening, the eggs or hatching young had been destroyed. In the deserted nest cavity I found only a few eggshells, along with the usual debris of a red ant nest.

5

Barred Owl Talking

TALKING OWLS SEEM A CONTRADICTION. OWLS DON'T TALK. According to folklore they hoot, but rarely, and nowadays most people haven't even heard a hoot. But I've heard them here in the Maine woods since I was a boy, and up on my hill near my cabin for thirty years. However, they don't all hoot. The saw-whet owl emits a high-pitched, one-second-long whistle, monotonously repeating it without apparent variation at about half-second intervals for hours on end. This tiny bird sounds like a city garbage truck backing up. Hearing this song induced me to put up nest boxes in spruce woods in the hope of attracting a pair to nest there. None did. I heard a great horned owl's deeply resonating *who-who-who-whooo* after midnights in the winter of 2007, when a pair nested in a pine grove next to my cabin. I was not closely acquainted with this pair, although several years earlier a great horned owl, which I called Bubo, had made its home here for at least two years, and I hoped it had returned.

Owls are almost invisible to us, and their presence is easy to miss, except when they happen to be vocal. That winter a pair took over the nest my ravens had made and used the previous year. In March, when temperatures had dipped to −29°F, I heard the ravens making a commotion at the nest site. Rushing out to investigate, I saw a great horned owl flying away from their nest tree. As I climbed the tall pine toward the nest, to my great surprise a second owl flew off, hooting at me. The owls' two eggs lay in a snow cavity; either they had scooped out a hole in the deep snow covering the raven nest or falling snow had piled up around the incubating parent. In other years I found evidence of the owls' presence: a half-eaten grouse stashed in a crotch of a large sugar maple, the remains of a grouse kill in the snow. One year the owls killed at least two just-fledged young from the nest of Goliath and Whitefeather, a pair of ravens whose family I had been observing. As recently as 2010, my neighbor Dionel Witham talked of "the big owl with ears" that perched in a pine tree on his property. I suspected it was Bubo, but I had no way of knowing.

During all these years I never heard or saw a barred owl, perhaps because great horned owls kill them.

Owls are vocal mainly just before nesting time, which starts in midwinter for the great horned owl and a month or two later for the barred owl. The barred owl's standard vocal signature, *who-cooks-for-you,* is well known. Much less familiar are its unearthly-sounding shrieks, like cackling, maniacal laughter. This caterwauling in the middle of otherwise silent nights has been compared to a crazy screaming. Wildcats or bears may come to mind as other possible sources of these sounds. Such a vocal display does not, however, seem to our ears an inducement to courtship.

From the total lack of owl sounds of any kind, I did not suspect that there were any owls at all in the woods around my cabin in the winters of 2011–2013. But then, during the night of April 3, 2013, I heard an "expletive hoot" (as I described it then). An owl was near! Of that there was no doubt. Given the woodland habitat and the vastly different calls of the saw-whet owl, it could only be a barred or a great horned owl. On several occasions I had heard blue jays scolding in the woods at what I thought might be an owl, but when I went to investigate, they stopped their commotion and I saw nothing, except once some freshly plucked blue jay feathers on the snow.

I wondered how an owl could survive in my woods at this time of the winter, when the snow was deep. How could it catch mice or shrews hidden under the snow? But proof of an owl's presence came a day after the expletive hoot. On my way to get water from the well, a mere hundred meters from the cabin, I found an owl pellet on the snow under a maple tree. I could not remember ever seeing an owl pellet so large: 7.5 centimeters long and 3.0 centimeters wide. Was it from a great horned owl?

Curious about what this owl had eaten, I dissected the pellet, which to my surprise contained the front ends of five skulls of the short-tailed shrew, *Blarina brevicauda*, and their matching jawbones. The owl had swallowed five of the mouse-sized shrews. I could not be certain it had eaten anything other than their heads, though, because I saw no other bones. It looked as if this owl had a fastidious taste for brains, and also a systematic killing method, since each skull's brain case had been crushed or obliterated.

Blarina are stocky shrews that are ten centimeters long without the tail. Like moles, they have no noticeable eyes and live in the soil or at least under the leaf litter, which was now covered by a foot of hard-packed snow. How could the owl have managed to catch even one of these shrews, much less five of them, in an obviously short time? I could think of only one place where this could have happened: at the birdfeeder stocked with black sunflower seeds that hung in the birch tree by my window.

Chickadees, nuthatches, and finches spilled seeds onto the snow, and I had seen the burrows and tracks of small mammals that had come up out of the snow to feed on the seeds. The owl could have perched here every night, and I would not have known! Was it Bubo?

It was too late in the year for great horned owls, and possibly also barred owls, to nest, but perhaps if one found a suitable home site it would remember and come back the next year. So on that same day I made a big bird box out of a hollow apple tree that I had saved when making my winter wood supply. I sawed a short cross-section of the trunk, notched a hole at the top, and attached pieces of board at the top and bottom plus one at the side with which I could nail the prospective owl house to a tree. I nailed it about six meters up in a red maple near where I had found the pellet. Its entrance was large enough to accommodate either of the two owl species.

Four months passed before I again heard any owl, and it was then not just one owl but two calling back and forth. It happened on August 29, which I came to refer to as "the night of the barred owls." I heard the barreds' unmistakable *who-cooks-for-you*, and it came almost directly from where I had attached the owl box. Another long session of back-and-forth calling, this one for an hour, occurred near dawn. There seemed to be excitement in the calls. Heavy rain and wind were coming from the east, and wind-tossed leaves were flying along the ground. Even an owl could not have heard animals rustling on the forest floor or seen their movements, so this pair's excitement was probably not from catching mice.

The next night, around 1 a.m., I heard the owl right next to the cabin. It must have been facing me, because the sound was quite loud. This time it wasn't the usual *who-cooks-for-you*, but a single very drawn-out *whooo* with an inflection at the end. Just one call — and about two minutes later, another. A long silence followed, and then, from a distance, a series of three similar *whooo*s. Then silence again.

Why were the owls calling now, at the end of the summer, far beyond the nesting season? The timing did not fit with the standard notion of why and when owls vocalize: the "passionate throes of courtship." Two birds were interacting, and they were conveying information. But what was it? I had no clue. From then on I took detailed notes every time they called, describing what they said, in the hope of discerning some pattern from which to decode their meaning.

At first it seemed possible to make progress in the decoding, because I heard the owls every night. All the way through September and into October, the near owl, and sometimes also a distant one, called at dusk, sometimes in the middle of the night, sometimes even in the middle of the day, but usually most vigorously just before or at dawn.

For weeks there had been an apparent harangue between my (the near) owl and another to the north. Then, from November through January, only the near one called, and rarely. But by February, just when I expected courting to begin, every night was silent.

In all those months I heard the weird shrieking that supposedly sounds like a crazy woman screaming only once, at dusk on October 4. An owl had called briefly at 4 p.m. from close to the cabin, and another had answered from a distance. The answering one came closer, after which the caterwauling started. Was it a showdown of territorial rivals, or a reuniting of mates? I had no way to explain the singularity of this vocal exchange. But then, much else about the owls' conversations was singular as well.

Usually my owl was the initiator of vocal monologues or exchanges. On October 13 at dawn it made twenty-six calls in a row. Instead of the usual four-syllable *who-cooks-for-you*, each call consisted of two of these phrases strung together into an eight-syllable utterance. The owl had commonly used the eight-syllable pattern with the final syllable greatly prolonged into a long *whooo,* either rising or falling in pitch at the end, sometimes smooth and sometimes with a tremolo. But this time the closing inflection was absent. Four minutes after the twenty-sixth call, the distant owl, perhaps a kilometer to the north, answered with four similar calls. Then, in mid-afternoon in bright daylight, the near one called again, but only six times, and another answered immediately, but only briefly. The next dawn my owl made a call I had never heard before: a drawn-out multi-syllable screech that I described at the time as *arrr arrr eeeah-oooh* with a downward inflection at the end. On November 4 it made only two long *whooo* calls, and on the next night six. At 6 a.m. on November 21, from its usual place in the pines next to my cabin, it made twenty repetitions of the standard *who-cooks-for-you*, except that each four-syllable sequence was again double. The next morning at dawn it called ten times in the double *who-cooks-for you*. That afternoon the distant owl called, and about two hours later mine erupted with seven long one-syllable *whooo* calls.

After this I heard no owls for five weeks and thought that perhaps both of them had left the area. Little did I know that one owl not only was present but had probably been watching me around the cabin off and on for at least two years.

On January 2, 2014, I suddenly acquired ten new pairs of eyes. They belonged to ten University of Vermont students who had come from Vermont to take my winter ecology field course: Nikki Bauman, Mike Blouin, Michelle Brown, Kat Deely, Kyle Isherwood, Stephanie Juice, Holly Kreiner, Ali Kosiba, Maddy Morgan, and Andrea Urbano. As usual, every one of them saw something new that I had never seen, but it was Kyle who on our third day spied the barred owl. Perched about fifteen meters up in a pine tree behind the cabin, it just looked at him calmly. In late afternoon, when the whole class gathered by the tree, it was still in the same spot. As we gawked it peered down at us for a couple of minutes but soon, as if bored, looked off to the side.

We convened at the cabin that night as usual and cooked and enjoyed a supper of soup and fresh, delicious corn bread. Gathered around the warm stove and the hissing propane lamp, we

The barred owl's perch when it visited at the cabin. (Only one of the trees in the forest is shown here!)

swapped stories about our finds that day. For example, Mike had found bobcat and coyote tracks where we had heard ravens call at dawn. The owl was not mentioned until a student who had gone outside wearing a headlamp rushed back in and blurted, "The owl is *here!*"

We spilled out the door and looked up: there it was, perched almost directly above us. Even with all our chatter and with several lights now aimed directly at it, the owl didn't flinch. In fact, it didn't even look at us. It was looking down, seemingly undistracted, in the direction of the birdfeeder. I wondered: Was it looking for short-tailed shrews coming to feed on the spilled sunflower seeds?

I had seen barred owls in the woods before, but never one this tame. This one clearly had long familiarity with humans, probably mostly with me over the last year. It, rather than a great horned owl, was the likely producer of the owl pellet with five crushed *Blarina* skulls. I had not seen this owl before because, unlike the students, I go to bed immediately after dark — unless there is a good excuse to stay up. There was a good excuse now, and I hoped to continue to observe this owl into the future.

Five days later I heard it call from the pines before dawn. It made two long, drawn-out, high-pitched screeches that mellowed at the end. There was no response from the distance. And that would be all I heard from the owl for almost a month.

On February 5 I lay awake and at 10:45 p.m. heard the owl's *who-cooks-for-you* three times. The call was clear, and very close. Proof: the owl was still in the area. But I was unprepared when, at dawn the next morning, I stepped out the door in a snowstorm and saw it perched in daylight on the same tree, same branch, same spot on the branch, and facing the same direction as when the class and I had seen it deep in the night a month earlier. Seeming to watch something on the ground, it paid me scant if any attention as I walked past it to the outhouse, and a minute or two later it was gone. But I saw its wing prints on the snow under the tree.

The owl returned at dusk, and again it perched on the same spot.

My journal entries about the owl at that time were little more than descriptions of the vocalizations I heard at night. I had not deciphered anything about what the owls were saying, except perhaps that there was much they *did* say. Nevertheless, I was thrilled to have made the acquaintance of a wild owl in its natural habitat. But a week later, on February 13, 2014, there was more.

Another snowstorm was on the way, and at dawn the temperature hovered around –10°F. A gorgeous deep blue sky was fading to green, and a line of orange appeared on the eastern horizon through a latticed black silhouette of maple trees as an incredibly bright morning star shone to the southeast. I peeked outside, as I now did every morning, to see if the owl was there. I had not seen or heard it for a week, and I didn't see it now. I made a cup of coffee, reclined on the couch, and picked up the March issue of *Running Times,* which had an article about me by the writer Scott Douglas. I was amused or bemused to read that "the competitive record" from my 1957 and 1958 high school cross country seasons "is in Heinrich's first journal, a 3x7-inch notebook that's half running diary, half nature log." I had lent Douglas my journal, and he included a quotation from 1957 that was not about running: "Apr. 21 — Barred Owl eggs ready to hatch." (Later entries suggest I meant incubate, a three-syllable word not in my vocabulary at the time.) I could not recall how, more than fifty years before, I had known the barred owl's incubation schedule. But I delighted in the fact that in the present my barred owl was approximately a month from egg-laying time.

After high school, as a freshman in English at the University of Maine, I was forced to write weekly essays. Thankfully I remember only one of these efforts. It was about a pair of barred owls that I observed at their nest in an old basswood tree in the forest near Pease Pond, about twenty kilometers from where I live now. I found the nest during spring break, and was so entranced by all the sounds its occupants made that I hid in the woods and listened to them for several evenings in a row. If I had met a band of just-landed aliens from another planet, I could not have been more excited than I was about these owls. I now had something to write about, and I could not *not* write. I needed to record the owl details so I could savor them later, again and again. I had probably deserved the grades I'd received on my other essays, but I knew that this one rated at least a B. I was eager, for once, to get my paper back. Then the unthinkable happened. I was the only one in the class who did not receive his graded essay. When I meekly asked the instructor about my paper after class, he looked me straight in the eye, smiled, and said, "I lost it."

I would give a lot to have that essay now, to awaken memories of those evenings at the edge of the pond in spring with the just-returned black ducks quacking in the evening light and the owls later calling. But no memories could top what I saw less than an hour after reading my journal entry of April 21, 1957.

As mentioned, I was reclining on the couch. A movement caught my eye, I glanced to the left, and there, no more than three meters away, was the barred owl perched on a limb of the birch tree. I watched it for about twenty minutes. From time to time our eyes would lock, then its gaze would return to the ground. It seemed very alert at times, and I talked to it; it must have heard me but was relaxed. Once in a while I saw it almost twitch with excitement as it leaned over and focused on something below my windowsill. Suddenly it leaned farther over, spread its wings slightly, then again settled back and relaxed. It had to be watching an animal that came and went. I got up to grab my camera.

I soon got to use it. Once again the owl tensed and leaned over, but this time it turned around on its perch, then wham-dived onto the snow and came up with a *Blarina* in its talons. In a few swift wing strokes it flew up to another birch, where it chugged down the shrew whole in a series of gulps.

Hoping to keep the owl around, I left a dead red squirrel on the snow below its perch, but night after night my offering remained untouched. On February 18, though, after a heavy overnight snowfall, an oblong groove in the snow showed that something big had plowed into the end of a set of deer mouse tracks; no tracks led away. Since the dead squirrel had not interested the owl but something moving had, I wondered if it detected potential prey by hearing, seeing movement, sensing heat (infrared, as some snakes do), or recognizing form.

The owl's hunting pattern, so readily available for experimentation here, might give me the opportunity to investigate what cues it used to find prey. So now I had to catch rodents to feed my friend. The way to its brain would be through its stomach. Barred owls are perhaps unusually catholic in their diet and ways of obtaining food. Although mostly feeding on voles, mice, shrews, and birds, they have been reported to chase amphibians on the ground, wade into water to catch crayfish, and dive to capture fish. These prey animals all have one thing in common: movement. Movement, it seemed, might be the best if not the only way for a generalist owl to survive over a wide range of habitat used by a diversity of potential prey. But there were other possibilities. I would go first for the low-hanging fruit.

On February 22 my owl was on its usual perch on the white birch. I brought out of cold storage a dead short-tailed shrew I'd been saving for it. These *Blarina*, I knew by now, were acceptable game. I threw it onto the crusty snow to almost the precise spot where four days earlier the owl had successfully intercepted a mouse. Knowing it could see that spot from its perch, I watched closely. The shrew was black, clearly visible against the white snow. Would the owl pounce? I waited ten minutes, but no, it did not budge. I recalled that rattlesnakes find mice in the total darkness of their burrows by "seeing" their body heat; the snakes have infra-red detectors (although not in their eyes). Might the owl also detect heat? I doubted it, but my doubt was irrelevant. Experiments and experience decide. So I retrieved the shrew, heated it on my wood stove, and returned it to the snow. Nothing happened. The owl sat still. Again I waited ten minutes, and again there was no response. So body heat was not the owl's hunting cue. Having also eliminated recognition of form, I deduced that movement and/or the results of it, such as sound, were the most likely cues used by this owl.

I had confirmation that movement was a primary cue from a long-eared owl I lived with while a graduate student at UCLA. The owl perched unobtrusively all day and part of the night on the almost ceiling-high corner of a bookcase in the small apartment my new wife and I shared. An unsuspecting grad student neighbor came to visit. While sitting on our couch he absent-mindedly twirled his wristwatch on his index finger, and the owl struck his hand. To say he was taken aback is an understatement.

A naturalist and prizewinning writer told me that during the winter a barred owl in his rural Adirondack setting fed on chicken meat he put out to attract birds. It would have been surprising if a hungry barred owl had *not* shown up at his bait. However, it probably had not initially come to eat chicken. Meat of many kinds attracts shrews, voles, deer mice, and red and flying squirrels. Any of these animals running around would be manna for an owl. But this is not to say that owls do not eat chicken, either dead or alive. Learning is part of their foraging tool kit as well.

Thread is a great tool for making a dead shrew move. With one end of a white thread tied to the shrew and the other end in my hand, I threw the shrew onto the snow, then pulled it along. Instantly the owl hunched over, launched, and came silent as a shadow directly toward the shrew (and me). It "caught" the prey almost at my feet, and in the light of my headlamp, as it looked up at me briefly, its eyes shone red. I pulled on the thread and the owl lifted off with the shrew in its talons. The thread broke, and after a flurry of silent wing beats the owl landed on the perch it had come from and gulped down its prize.

At dawn the next morning the owl was back at the same spot.

It continued to come many nights throughout the winter, and sometimes even on the brightest sunny days. To me it became like a pet, but with the advantage that I didn't have to provide for it. On occasion, though, I eagerly did so, and it eventually began to take meat from the ground next to my feet. Often it came flying out of the woods to land on its spot on the birch as soon as I opened the cabin door. It had learned where and when to watch for me.

March 31, 2014. A big temperature drop and a deposit of hard-driven snow from a northeaster had purple finches, goldfinches, mourning doves, evening grosbeaks, the first returning junco, and at least ten blue jays crowding the feeders. The crows were in pairs now, and I watched one carry a twig to a partly built nest in the crown of a spruce. The owl seemed to have left — I had seen no sign of it for weeks.

April 1, 2014. A gorgeous clear and cold day. The snow crust was strong enough to walk on for the first time this year. There was no sign of a raven nest, but I saw a single raven fly to the nest site three times. The pair of crows landed near a roadkill deer, cawed, and left. A raven flew over. But the big event was at night. I awoke near 1 a.m. to caterwauling, opened the window by my bed to hear it more clearly, and listened to the barred owl calls I had not heard since fall, the familiar four-syllable sequence *who-cooks-for-you* (henceforth abbreviated as wcfu). Consecutive sequences were evenly spaced, at about one every half-minute.

The owl continued with scarcely a pause until daybreak, with mostly the usual wcfu but sometimes the eight-syllable sequence I had heard the year before, made by doubling the wcfu and drawing out the last *u* into a long *whooo* with a descending tremulous pitch trailing off at the end. There were long pauses between these songs, but I kept listening. At around 4 a.m. the owl called from near the back of the cabin and another responded from the north trail, one wcfu almost instantly answered with another. The owls were answering each other so quickly that they sounded like just one owl calling. It seems that at least some of the wcfuwcfu sequences that I had always assumed were by one owl were instead duets! With this in mind, I listened for a nuance that I had not been conscious of before, and I got up and went outside to see if my owl was on its perch on the birch tree; it was not.

Spring and summer came on schedule, with frequent hooting as well as weird cackling and screeches. Fall arrived in its glory of color, the leaves fell, snow came, and then there was silence. As the season slipped into the third winter since the owl had first arrived, there was no sign or sound of it. I missed it, and when outside in the evening or at night, I almost always glanced up or shone my flashlight to its perch on the white birch. The long nights got longer and the short days shorter, and I wondered if the owl would ever return. Perhaps it had lost its mate and moved on. It might have died; owls are frequent roadkill. Some years ago while driving through New Hampshire one night, I passed a seemingly dead barred owl, then in my rearview mirror saw what I thought was movement of a wing. I turned around, retrieved the owl, and put it on the back seat of my pickup truck. Minutes later it revived and perched next to me on the backrest of the passenger seat, where it revived fully. A rare occurrence.

Finally the solstice arrived, with its promise of nature's reawakening. I had in the meantime had my own good fortune, having found a partner who loved the woods as I did, in spring, summer, and fall, as well as winter. The snow was deep by then, and my partner and I decided to celebrate in a fitting way: with a big bonfire and toasts of red wine by its warm glow. I chainsawed dead dry branches for the blaze plus some green ones for sparks, piled them up, and with a match set it all off. A large flame shot up into the moonless night.

We added green branches to the hot coals, and sparks flew up in a stream. As we drank a toast my eyes followed their incandescence upward, and there against the black sky I saw, highlighted by the fire below, the gray-white outline of a huge owl just leaving. With several beats of its great wings it flew over the cabin and into the dark woods. But it came back after the flames subsided. I tossed a slice of the venison we were having for supper onto the snow directly in front of us. The owl leaned over on its since long-used perch and with no further hesitation swooped by us to pick up the food.

Surely it was the owl I knew, my friend from the year before, the one that had learned that what I tossed for it was food. It did not need to see fur or feathers, provided the meal came from me. The next day in the light, it swooped down and grabbed meat out of my hand. It continued to perch on the same large birch tree, the same branch, and the same spot on that branch where I had seen it countless times before. These behaviors could not have spoken more clearly who it was.

In the moment of joy and mystery when I realized that this was "my" owl, which was back and would again be here for weeks, months, or maybe longer, I felt connected with all the moments of my past and now my prospects for the future.

6

Hawk Tablecloths

A PAIR OF BROAD-WINGED HAWKS, *BUTEO PLATYPTERUS*, comes every year from South America to nest in broadleaf forest adjacent to our Maine cabin. In 2011 the pair arrived on or near April 25, when I saw one of them perched in a black cherry tree at the edge of a hole I had dug years earlier. The hole fills with meltwater in the spring. This vernal pool has become a breeding place for wood frogs, spring peepers, and salamanders, and later in the summer it is also populated by leopard and green frogs, dytiscid and gyrinid water beetles, and of course larvae of dragonflies and mosquitoes. The hawk was hunting wood frogs, which, just as the last ice is melting in a warm rain, emerge from under the fallen leaves and travel to their breeding pools to mate and spawn.

Broad-winged hawks return later than other local hawks, probably because they wait until frogs and snakes come out of hibernation. As the name "broad-winged" suggests, these hawks are not built for swift pursuit, but snakes and frogs are suited to a sit-and-wait hunting strategy.

They also nest late. I had in the past climbed to photograph their nests and taken pictures of clutches of two or three eggs with their typical spots and blotches of chocolate, purplish, sienna, tan, and various other shades of brown. The eggs are a feast to the eye on their nest molds, which are always chips and flakes of tree bark.

It seemed odd for a bird to use hard bark chips, let alone to favor beech bark chips to line the nest. Live beech tree trunks have a smooth, solid surface, so the hawks can get the chips only from dead trees with flaking bark.

For several years the broad-winged hawk nests were located a few hundred meters down the slope from my cabin, in a tall sugar maple tree where it divides into a quadruple crotch. Soon after seeing one of the hawks at the vernal pool, or more often the pair circling in the blue late-April sky, I'd check to see if they would reuse the old nest or build another nearby. The old nest had not been used in the last two years. But now in 2011 I saw signs of activity around it again. The best clues were two or three ash twigs on the ground beneath the nest that showed light fresh wood at one end, indicating that they had recently been broken off a tree. The hawks were rebuilding their old nest.

On May 21, when the beech leaves had been unfurled for several days and the matted dark brown leaves on the forest floor were sprinkled with gorgeous blue violets, purple trilliums, and white star flowers, I saw a hawk on the nest. From the rise of the hill where I viewed it through binoculars, it seemed directly in front of me, peering over the nest edge, as I looked into its yellow eyes.

Almost a month later, on June 16, she (the presumed female) was still on the nest. I had no particular inclination to climb to the nest; I had visited others before and the rewards now seemed less than the effort. But then company arrived from Cape Town, South Africa, and changed the equation.

Greg Fell and Dean Leslie, two jovial visiting filmmakers from

The African Attachment working for Salomon Running TV S03 E01, came to make a short film about my lifelong interest in running (it is available on YouTube under the title "Why We Run"). When they wanted to record me doing something outdoorsy other than running, I suggested that I climb a tree and proposed the maple with the hawk nest. If lucky I might see the baby hawks; I had never been close to a broad-wing nest with young. The filmmakers strapped a small video camera to my head so I had my hands free to climb.

As I began to climb, the hawk sitting on the nest flew to another maple tree. It called several times but did not seem agitated even when I reached the nest. The large stick nest held one egg and one pure white downy chick that was surely the cutest baby bird ever hatched from an egg (it can be seen in the video clip). Under it lay — to my surprise — pristine fresh fern fronds. The baby cheeped feebly, as though begging for food. The camera recorded this, and my new friends from South Africa were pleased. To them the nest lining of fern meant nothing, but to me, seeing the ferns instead of the expected bark chips was everything, because I had never before encountered this way of lining a nest.

The whole nest mold was lined with a huge fern frond that must have been placed there an hour or two before my climb. Under it lay another frond, slightly wilted, that had probably been there for a day or so. I was amazed and baffled to see fresh ferns, because the nest "should" have been totally finished a month earlier, before the eggs were laid, and with the usual dry tree bark chips as a foundation for the eggs. This was a puzzle, and I had an opportunity to solve it. I knew I would make more visits to this nest.

I climbed to the nest nine times in the next thirty-five days, until the young fledged on July 22. The parents replenished the greens almost daily for a total of fifty-five green sprigs, averaging two new sprigs per day in the first three weeks. At first the greens were ferns, but on June 26 the nest was lined with matted-down and wilting green sugar maple leaves with another very fresh fern frond on top. On July 4 the whole nest mold was solidly lined with fresh green cedar sprigs and a sugar maple twig with four unwilted leaves that could not have been a day old. It was 2 p.m. and 80°F; the greens must have been brought in that morning. Unlike any other birds I knew, these hawks were continuing to "build" their nest after the young hatched and even as they grew, almost until fledging.

Lining the nest with greens must serve a purpose (confer an advantage), since it involves a cost. In order to get the ferns, for example, the hawks had to collect them from the ground, and getting the cedar sprigs required trips to a grove at least a half-kilometer away. The composition of the nearest trees did not match the nest contents. Therefore, the birds had exerted choice.

I thought I could eliminate several alternative hypotheses of what the greens might do. Greens would not camouflage white nestlings. They could not have been nuptial gifts for a mate since there was no more mating or mate choice during the nestling stage. Cushioning and insulation were also not likely explanations, because the materials were neither fluffy nor soft and if serving as cushioning should not have been laid over solid bark chips. Nor would they have been incorporated this late in the nesting season. Improving the nest structure for future use was excluded because the lining would retain moisture and hasten rather than retard decay. Aromatic medicinal plants, used to possibly kill parasites, another potential option, also seemed unlikely. To be useful, such plants should have been incorporated *into* rather than placed *onto* the nest, and in any case the hawks' choice of non-aromatic maple leaves and rejection of readily available conifers such as pine, spruce, and fir would exclude that hypothesis as well.

But there is often more than one advantage to any given act.

The one consistent aspect of the greens the hawks chose was that they provided a flat, clean surface. Spreading a fresh layer of greens may be analogous to our spreading a clean tablecloth (if we ate our meals without using plates). Hawks often store surplus food in the nest; on one of my inspections this nest contained a fresh, half-eaten young woodcock as well as remains of red squirrel and grouse.

Since these hawks nest late into the summer and bring meat into the nest in the hottest part of the year, fresh greens as lining could serve a hygienic function. A clean substrate would reduce the accumulation of bacteria and hence retard spoilage.

A year later the pair built their nest in the fork of an ash tree, and in the subsequent winter friends and I built a blind in a large red maple twenty meters upslope of the ash to be able to see into the nest if the hawks reused it the following spring. The hawks returned and again hunted frogs at the vernal pool, but they did not use the nest where we had built the blind. Most gambles are lost, but they often generate observations that can provide opportunities for other investigations. The next time led to blue-headed vireos.

7

Vireo Birth Control

WE EXPERIENCED THE COLDEST NIGHT SINCE SPRING ON September 11, 2011. Temperatures dipped to 40°F. There was not a breath of wind under the clear blue sky. The red maple leaves were already half turned and most of the wild-growing apples had fallen from their trees in the woods. Robins, thrushes, and a flicker had picked all the chokecherries from the clearing, and most of the migrant birds were gone. The almost deathly quiet, compared with the raucous summer clamor, was deafening — until suddenly, around 8 a.m., I heard the slow, languid, clearly enunciated syllables of a blue-headed vireo in full song. The call did not come from the red spruces, balsam firs, and white pines alongside the clearing where the blue-headed vireos had lived in the spring and summer. Instead, this bird was singing from a lone white birch at the edge of the clearing near where a red-eyed vireo pair had nested in a sugar maple that summer. I had to look closely for its distinctive white eye ring to convince myself of what I was hearing.

He (female songbirds do not sing) sang in the birch for a minute, then flew into the conifers in the nearby woods and sang for another half-minute, and then was silent.

The brilliant burst of song made an impression because of the bird's unusual visit to the clearing and the unusual time of year: I had not heard a blue-headed vireo sing since the spring. I immediately retrieved my sketch of the bird from the spring and made improvements. This was an extraordinary event, and I wondered if the song was a response to something significant for this bird. Why now, why here, why so loudly and emphatically, and why so briefly? Why did it feel like singing? Birds were migrating; perhaps this one was passing through and was staking it out as a possible nesting site for the next spring?

April 17, 2012. The ground in the woods was still under snow, but the first yellow-rumped warblers were returning, singing their lisping songs and foraging in the tops of the now blooming red maple trees. Ruby-crowned kinglets were migrating through and occasionally breaking into their staccato song-chatter on this windy and 30°F day. But what made the day for me was a blue-headed vireo that arrived and spent most of his time at what I thought would be an ideal nesting place. All morning long he sang continuously, loud and clear, circling all around the clearing, and by mid-afternoon he ended up where I had heard him or another blue-headed vireo sing so beautifully the previous fall.

After this I was pleased to hear his songs of short slurry whistles with their upward- and downward-inflected notes daily.

Two vireo species live here on the hill. The red-eyed lives strictly in deciduous trees and does not return until late May, well after the trees have leafed out. Its loud and energetic song is a constant all summer until August. The blue-headed vireo comes back a full month earlier than the red-eyed, long before any broadleaf trees have started to break bud. Here in Maine it lives and nests in conifers. Its song is similar to the red-eyed's except that it is slower-paced and is heard when there is still snow under the spruces and firs. The blue-headed vireo normally quits singing in early summer. But birds of any given species don't always abide by the norm. I had seen blue-headed vireos nest low only in conifers, but in 2012 I found a nest in an unusually placed (most are in clearings) tall chokecherry bush near the sugar maple tree where the broad-winged hawks had nested. The outside surface of this nest was white from fluffy raptor down, rather than the paper from hornet nests usually found on both red-eyed and blue-headed vireo nests.

One often comes upon the unusual serendipitously, by play, by involvement with something entirely different. At least that is how I came to see a case of apparent avian birth control.

On May 11, still before the trees had leafed out, while following a sapsucker through the woods, I heard and then saw a pair of blue-headed vireos. They were keeping in touch by calling softly to each other. One of them flew to the trunk of a white birch, tore off a thin strip of bark, and flew off carrying the strip in its bill. Its mate followed, and so did I, hoping to find the nest. The birds were soon out of sight, and I searched in vain for their nest at every spruce and fir tree I passed. The long strand of bark was probably for use in the beginning of nest construction; according to R. D. James, the unmated male chooses a nest site and starts building the nest by looping strands of material over a horizontal forked twig. The female indicates acceptance of him by helping, and she adds the nest lining on the final two of the eight days of nest building.

I encountered the pair in the same area two days later. Again one of them tore a thin strip of bark from the same white birch, then flew off with it in the same direction as before. This time when I followed them I found the nest. To my surprise it already looked fully formed, hanging, like all vireo nests, in the fork of a twig. It was not in a conifer as I had expected, but instead about four meters from the ground in the aforementioned tall choke-cherry bush. This was the first time I had ever seen a blue-headed vireo nest in a deciduous bush or tree.

A second blue-headed vireo nest was not far from this one, and I found it too by following a pair. One or both birds made short *purr* sounds that seemed like whispers in and around an area of dense conifers. I often lost sight of them, then sometimes saw only one that perched and preened in the trees. Their nest was suspended near the end of a branch in a large balsam fir tree, about six meters above the ground. Sensing a possible story, I wanted to photograph these vireos feeding their young. Since they would not come close to me, I had to get close to them. Over the next several days I built a platform, attached a hide in the form of a little hut to it, and placed it in the tree next to the nest so my 18–55mm lens could reach it.

Throughout my construction of the bird blind, one of the vireo pair was almost always incubating on the nest. It did not flush or seem concerned by my presence. On several occasions it sang next to me; I had expected alarm calls but never once heard them. And yet these vireos were not inattentive. One day a male blackburnian warbler took a bath in a puddle near the nest tree, then flew up to a branch and started to preen. One of the vireos swooped on him and chased him off, making a strikingly scratchy noise all the while.

Later on I fared no better.

When I settled into the blind to start photographing, one or both vireos scolded me in loud raspy and scratchy chittering that reminded me of fingernails scraping across a solid surface. One picture of the pink-yellow naked babies hunkering down was all I got, and was happy to get, before two days of heavy rain (May 29–30) curtailed my efforts. Then came one gorgeous day followed by three more days and nights of torrential rain. The nestlings somehow survived, but the parents remained too wild for me to photograph them, so I switched my attention to the atypical nest in the chokecherry bush.

The bush was too thin to climb, but the nest was low enough to look into from a four-meter stepladder. To let the birds get used to the ladder, I first placed it about ten meters from the nest, then moved it closer each day for several days. The incubating bird sitting on the nest never budged, even when the ladder was right at the nest. When I climbed up to photograph the four eggs, it hopped off briefly, then settled back into place and ignored me. The four eggs were mostly white and were about ready to hatch. (When freshly laid, the eggs are colorless and have thin, pearly transparent shells that later look chalky.)

On May 31, hoping to see young, I again very slowly climbed the ladder to the nest, talking to the incubating bird softly in the hope that it would understand that I didn't intend to do harm. It now remained still even as I got my camera lens within a half-meter of it.

Four days later, after three all-day and all-night torrential rains, I went back expecting to see either young birds or an abandoned nest. Instead I again found eggs in the nest — but now only three, not four. An egg could not have disappeared on its own from the vireos' deep nest cup, and a raiding predator would not have daintily plucked out just one egg and otherwise left the nest in pristine condition.

Two days later, on June 6, I checked the nest again. A bird was sitting on it as before. I talked to "her" as before, and she swiveled her head in my direction. Even when I leaned directly over the nest to take pictures she didn't budge. Deciding to try for a close-up, I pulled the branch with the nest toward me and held it against the ladder. She still showed no sign of alarm. On an impulse I reached over and put my fingers under her belly. When she did not object, I gently lifted her to expose the eggs and with my other hand took a picture of her, the nest and eggs, and my fingers under

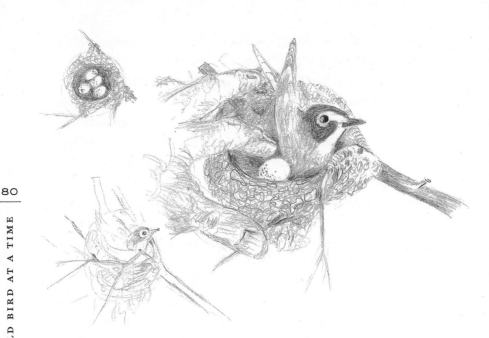

A blue-headed vireo on its nest, with the author's finger reaching under its belly.

her. At last she stood up, hopped out of the nest, landed on a twig next to me, and sang! So it seemed the bird was not a female as I had presumed; it was more likely a male, as its song was typically sung by blue-headed vireo males.

I got another surprise when I looked into the nest: it now contained only *two* eggs. Another egg had disappeared.

I found the nest abandoned a day or so later. The pair's departure was probably not due to direct soaking by the rain, because vireo nest-lining is coarse, apparently built to allow water to drain through it without collecting in its open-cup structure.

In trying to fathom what had happened I first examined the two remaining eggs. Each contained a decaying embryo. One of them had been dead for at least three days, but the other had died earlier.

The seemingly unending cold, wet weather was the most likely killer. Insects had become scarce. One of the parents might have been killed, as I had not recently seen two birds at this nest.

And two eggs had been removed. I had seen a similar egg reduction before: in a phoebe nest that was being incubated during a freeze in 2011. On March 23 it held one egg. Three frosty days later there was still only one egg when four would have been typical, but on March 29 there were three. I expected incubation to start then, but instead, when I next checked the phoebes' deep nest cup, the number of eggs had been reduced to two. The parents incubated these two, and two young later fledged from that nest.

Had the vireos voluntarily reduced the number of eggs to match the dwindling food supply? The number of eggs various species lay is affected by the availability of food; fewer eggs are laid when it becomes scarce. When the food supply diminishes during the nestling stage, the number of young fledged is often lower than the number of eggs laid because of starvation or siblicide. If an extra egg, such as a cuckoo's egg, is added, it may be tossed out. But I had not known of any birds exercising control over the number of their own eggs after laying. In this vireo nest, had all of the eggs hatched and the unexpected bad weather continued, the death of all the young would have been almost certain. But, by halving the brood, the parents might give at least some a chance to live. Birth control is practiced in any case, voluntarily or involuntarily. To date, parental egg reduction has not been proven. It would be incredibly difficult and require astronomical luck to be observing at the moment when a wild bird removes one of its eggs from the nest, and it would be morally indefensible to try to test the hypothesis experimentally. However, brood reduction is certainly more efficient economically than infanticide, which takes place after the wasted effort of incubating and feeding the doomed offspring.

The usual methods of reducing brood size to accommodate food supply are more cruel and less efficient. In eagles and herons one of the young eats the others or shoves them out of the nest. At the same time the vireos were reducing their number of eggs, in other nests I was watching, four tree swallow babies (two of which almost made it to the fledging stage) and five red-breasted nuthatch chicks died of starvation.

8

≋

Nuthatch Homemaking

IN FEBRUARY OF 2009, I ROUTINELY NOTICED A PAIR OF red-breasted nuthatches near my back door in Vermont. They typically inhabit dense spruce forests. Here there were just a few white pines, but the nuthatches probably stayed because of the constantly available sunflower seeds in my feeder.

On March 14, while there was still much snow on the ground and the beaver pond was still frozen over, the pair was examining dead tree stumps, and for a few minutes I heard a soft and steady tapping on a poplar stump. Unlike woodpeckers, which tap on trees to find grubs and to excavate a home site, red-breasted nuthatches hammer at wood to make a nest cavity, in which they also build a nest. So it was clear that these two were getting interested in making a home, testing various sites. Later that same day they started hammering near the top of a recently dead red maple tree trunk. The wood was hard, but they stayed, evidently preferring it to the softer poplar they had just tried. Progress was slow. Seven days later they had managed to excavate only an entrance hole. When either of them went in to work on deepening the hole, the end of its tail stuck out.

The red-breasted nuthatch, a tiny bird weighing only about as much as two nickels, tackles a challenging homemaking task. Unlike the local white-breasted and the European nuthatch species, which seek out and use existing shelters, this nuthatch hammers out a nest cavity in solid wood. The European species reuses existing shelters such as old woodpecker cavities and refits the too-large entrance holes by cementing them partially shut with soft mud that dries in place. The white-breasted nuthatch also uses preexisting cavities, but it does not reduce the size of their entrances. The homemaking progression of these three species resembles that of humans possibly first choosing a cave, then reducing the size of the cave entrance with logs and stones, to make

Portraits of a red-breasted nuthatch pair.

the cave into an independent dwelling that can be furnished for comfort on the inside.

To excavate an entire nest hole in solid wood, the tiny red-breasted nuthatch has to start early. "Early" is relative to everything else in the flow of the season. By March 27, after almost two weeks of labor, when the snow was melting and the ice on the beaver pond had a ring of water around the edge, only the tip of a tail showed when the bird worked. I recognized the female (by her paler head and chest coloration) as the one doing the heavy work. She hammered in four- or five-second bursts, backed out of the hole tail first with wood chips in her bill, perched at the entrance to fling them to the side, then slipped back down into the hole and hammered some more.

The male stayed in the vicinity and occasionally made a series of long, nasal calls that sounded like *aank*. Now and then he flew to the nest hole, peeked in, and chattered to her in apparent excitement, but he never set a foot or wing inside the entrance. When she was not excavating, he sometimes *aank*ed continuously for ten minutes at a time. Apparently he was calling her, because when she arrived and entered the hole to resume work he immediately became silent and then left. He seemed to have an active interest in the work, but not in doing it.

Snow fell for two days in the first week of April, then the air warmed. Maple sap gushed into my buckets, and I boiled it until late at night. In the mornings I got a roaring fire going under the evaporator and soon had plumes of steam rising into the cool air.

Waiting for water to evaporate is not usually exciting, but I was not only harvesting syrup; I was gathering other riches as well. The sapsuckers were back: I heard them drum in the woods. I saw a Canada goose sitting on her nest on the beaver lodge at the precise spot where she and her gander had landed on the day they returned (March 13) after their ten-month absence. She was probably already laying eggs. I marked on my calendar that the young would hatch around May 10.

In the afternoon of April 8, the nuthatches were again re-peatedly at the nest hole. Again he only looked in and chattered excitedly. But this time, after she went in, there was no more ham-mering inside or wood chips thrown out. She stayed inside for ten minutes. I suspected that, after about twenty-eight days of work, the nest site was finished. The amount of effort seemed extraor-dinary: a robin can build its nest in a couple of days, and some birds build hardly any nest at all. These nuthatches had produced a site for the nest, but after nearly a month of work they still had to build it.

Three days later, on the gorgeous clear morning of April 11, the goose was incubating. Mallards were quacking, and wood frogs chorused in the evening after the thin sheets of ice formed over the pond the night before had melted. The female nuthatch made one trip after another to a high branch in an ash tree, where she plucked bits of moss and then dove down off the tree directly to the nest. She chittered excitedly each time she slipped into the cav-ity. In two or three seconds she emerged and went back to the ash. The male, as usual, stayed in the background and *aank*ed. But occasionally, when she was not inside, he flew to the nest entrance and looked in. At these moments she perched on a twig nearby and quivered her wings as though mimicking a baby bird begging for food, but I did not see him react to her begging.

But two days later, on April 13, when the weather had been cold and the pond had iced over once more, he did arrive at the nest hole with his bill full of grass and what looked like shredded cedar bark. But unlike her, he did not enter the nest. Instead, he perched at the entrance hole, dropped the material inside, and then perched upside down, and after a while poked his head back into the hole for a peek.

She no longer brought anything to the nest, and she stayed in it while he occasionally passed or dropped small items to her. Some-times she poked her head out and I heard soft whispering calls or conversations between the two. These were, apparently, preliminaries to mating.

On April 20, at dawn, he flew to the nest hole several times in succession and twittered there, while she stayed inside. Sometimes he left, came back in a few minutes with something in his bill that he passed to her, lingered at the entrance twittering softly, then hopped onto a nearby branch and seemed to wait. Again he went to the entrance, appeared to talk with her, and flew back to the same perch to wait once more. Finally at 6:30 a.m., after he had done this several times, she suddenly shot out of the nest hole and perched on a twig a meter from it, where he joined her and they mated — all within one second. They repeated almost the same pre-mating and mating pattern on subsequent mornings.

There were now long periods when there was no activity at the nest.

By April 21 I had the impression that she was staying in the nest overnight and he was spending the nights outside. I heard him in the pine grove every morning before first light.

On the next morning I arrived in the dark and waited at the base of the tree. I did not see her enter the nest, and since she later came out she surely had spent the night there. He first visited her at dawn, peeking in and soft-chittering with her, then suddenly went on a nest-material-hunting binge, making seven trips between 6:26 and 6:57 a.m. Now, after twenty-eight days of house making and several days of nest making, it was surely time for egg laying. Given a normal clutch of six or seven eggs, one arriving each day, I expected incubation to start in a week.

By May 7 the female was staying inside the nest hole more or less continually, presumably incubating. I had constructed a platform of boards in an adjacent tree where I could perch to watch them from two meters away; they paid me no apparent attention. The male usually came from his overnight quarters in the pine grove at around 6 a.m. to chitter to her at the nest hole, and as always she responded with soft whisper sounds from the nest. Then he left, and sometimes she accompanied him briefly. She did not respond to his long *aank* calls in the distance. But whenever he came near the nest tree he signaled to her in a softer call, and she instantly responded with soft chittering calls from the nest, then popped up to the nest hole and peeked out, as he flew to her, and they engaged in an almost musical vocal exchange that would have been barely audible to me if I hadn't been right next to them. Sometimes when he arrived giving his invitation calls she flew out the hole and they met on a branch, where he offered her a morsel of food he had brought and she immediately swallowed it. Since the ever-hungry young would normally get feeding preference, I assumed that the eggs had not yet hatched. She never stayed outside the nest hole more than six minutes, but in that time she often foraged briefly, pecking at what may have been tiny insects on leaves and branches.

During the second week in May he continued to feed her what looked like crane flies and spiders. But apparently this diet was not enough to satisfy her appetite, because on her short periods away from the nest, now always with him, they flew to my feeder and took black sunflower seeds. They carried one seed at a time to a nearby maple tree and wedged it into a crack of the hard bark to hold it while they hammered it open.

When they returned to the nest tree, she often continued to forage in its vicinity. At these times he appeared anxious to get her to resume incubating. He indicated his impatience by perching on a twig a meter from the nest entrance and vibrating his wings and tail. If she did not immediately fly back to the nest, he flew to it, looked in, and again began to vibrate. At that point, with him giving the signal at the nest site itself, she invariably flew over, slipped inside, and resumed incubating the eggs.

I was away for six days at the end of May. During my absence there were three cold days with incessant rain, and temperatures at night dropped to near freezing. When I came back the bird-feeder was empty and the suet was gone.

I had looked forward to seeing the nuthatches feed their young. But instead, watching the nest on May 31 in my usual dawn session, I saw and heard nothing — absolutely no sign of a nuthatch. Then at 6:30 a.m., when I was ready to leave, I faintly heard the male calling from some distance. I stayed. He came to a nearby pine and called for thirty-six minutes without a single break, behavior I had never observed before. And I saw no sign of her. Finally she arrived, silent, at the nest hole and went in. After some moments I heard hammering from inside for a second or two, then quiet. Within thirty seconds she re-emerged, still silent, and flew away.

Her two-second bout of hammering would not have occurred during egg laying or incubation or in the presence of young. Perhaps it was a displacement activity, something nonfunctional that animals do when they are frustrated and don't know what else to do. I realized then that the nest had failed.

I never saw or heard either of the nuthatches again. To find out what had happened, I removed a slab of wood from the side of the log to reveal the nest inside. It was a mere ten centimeters down from the tiny entrance hole. Made mostly from soft cedar bark fibers and some fluffed grass, it did not contain hair or a single feather. Curiously, I saw no eggs and no young — not even the depression of a nest cup.

Knowing the nest could not have been empty all this time, I removed some of the nest material. And there, under the fluff, lay six tiny brown-spotted eggs. I opened one and found a partially decayed embryo.

Normally the eggs would not be covered unless the incubating bird expected to stay away for some time. This meant that the female, who had been the sole incubator, had apparently deliberately covered her eggs before leaving them. And she would have done that only if she had been very hungry and expected to be looking for food for an extended period. I concluded therefore that the birds had abandoned the nest because they had run out of food; he had been unable to feed her, and she had been forced to go into the forest to forage for herself.

Red-breasted nuthatches normally live in coniferous forest, but here they were out of their usual range. There were no spruce here, no fir, and the few pines now had no seed. Once my birdfeeder was empty, the only likely or possible food was insects, and they would not have been active during the three drenching and cold days.

For humans as for birds, making a home requires more than a house. These nuthatches had made themselves a place to live, but it was insufficient without a sustaining home environment. Like the sunflower seeds on which the nuthatches had staked their survival, our food supply is risky: it comes from very far away. But sunflower seeds at a feeder are perhaps more predictable than much other bird food can be.

Some years all the spruces in an area are loaded with seeds, or the oaks and beeches are bearing fruit in abundance, but in the next few years they may be barren. With no consistently predictable time or place to go, what then? What if you are a finch, a red-breasted nuthatch, or a blue jay and your seed crop is absent one year? You could move and trust your luck, or stay and make flexible decisions. To make good choices requires intelligence. Blue jays, members of the crow family, are thought to have that in spades.

9

Blue Jays in Touch

BLUE JAYS' INTELLIGENCE AND ABILITY TO TALK HAVE LONG been of interest. Mark Twain broached the topic about a century and a half ago in his famous short story "What Stumped the Blue Jays." The first sentence is assertive and authoritative: "Animals talk to each other, of course, there can be no question about that; but I suppose there are very few people who can understand them."

Twain's protagonist, Jim Baker, a "middle-aged, simple-hearted miner" from the mountains living "in a lonely corner of California," believes that blue jays talk to one another. He tells of a jay that started to fill a knothole in his cabin roof with acorns. Finding the task difficult, it recruited friends to help, and the flock proceeded to fill the entire cabin with acorns. Despite having reservations about the veracity of parts of this tale, I do believe that blue jays talk to one another. I just don't think they speak English.

· · ·

At dawn one day in late April 2007 I was seated and partially hidden in the branches of a pine tree. A pair of blue jays loitered nearby, making soft calls below me in still-bare viburnum bushes. The male approached his mate, and she began to imitate a baby bird, flapping her wings and opening her bill. He understood and placed food in her mouth. She then broke a dry twig off the bush and they both flew to the nest they were beginning to build in a small spruce in the forest. Six days later, while pulling rootlets out of the ground for the nest lining, they made soft, fast-repeated squeaky sounds. Their calls seemed like whispers, meant exclusively for each other, in stark contrast to the jays' usual exclamatory screams.

The jays' loud scream calls are given both singly and in sets of several in succession. Volume and nuances of pitch and rapidity of succession vary. Jays are commonly assumed to be scolding, like the other highly vocal familiar animal of the north woods, the red squirrel, which will chatter and stamp its feet on a branch above you if you enter its territory — not a welcoming display. The main difference, it seemed to me, was that the blue jay does not stamp its feet.

On hearing a distant jay's scream I often wondered what the bird was excited about. My guess was that the jay was agitated by an owl, a deer, a fox, or some other animal. I often tried to get confirmation, but although jays do mob owls, I was rarely successful in finding anything at all. However, deer and other woodland animals are both common and often difficult to see, so the hypothesis that the jay was announcing a woodland creature was difficult to disprove. The birds kept moving, too, so that by the time I got near the site of a scream both the jay and the supposed object of its screams were long gone, or there had not been any object in the first place. It was probably the latter: in many hundreds of times of watching a screaming jay I did not see any likely cause for alarm or comment near it more often than expected by random chance, nor was I ever scolded by a jay except at its nest. In contrast, I've been scolded by red squirrels on probably hundreds of occasions. If a human venturing near a jay is not scolded, why would a jay scold a deer, a moose, or any other large animal passing through the woods?

Having no idea what most of the blue jays' screams meant but suspecting they must mean something beyond attracting other jays to mob an owl or a hawk, I tried to get clues by looking systematically as occasions presented themselves. Although over the years I had taken copious notes of when and where blue jays called, I had found out little except that they were much more apt to call when perched than when flying. Also, they were overwhelmingly most likely to be seen singly, and secondly in pairs. To get some numbers to document that impression, I watched them both in my woods and on road trips. I recorded 168 sightings of singles. Next most common were pairs (32), followed by triples (10), and foursomes (4). Many more stationary birds were heard but not tallied.

My observation that blue jays usually called while perched and alone seemed odd. In most species of birds the crowds are often conspicuously and sometimes continuously noisy while in flight, whereas these jays were usually silent in flight. Ironically, on rare occasions (as I will describe in detail later) they aggregated into groups, and then they were either very noisy, or silent, depending on whether they were mostly perched or in flight. Why would they scream when they were alone, yet be noisy in crowds? The specifics and contexts of their vocalizations mattered, but which ones?

Unusual encounters with jays in the wild were opportunities for natural experiments, and one happened while I was pushing my way through a thicket of young alders and pines. I suddenly saw at near eye level a blue jay nest with a parent's tail sticking out at one end. The jay faced me point-blank, and screamed at me. In seconds its mate arrived and began scolding me as well. The racket had a result: within a minute six other jays gathered nearby in the tall poplars and maples. They started bobbing their heads and making both rattling and two-note, flutelike calls that contrasted with the excited screams of the pair next to me.

The varied vocalizations of the six new arrivals, however, did not sound like warnings. I suspect these newcomers saw that I was not one of the jays' archenemies, an owl or a hawk, that they might have expected when they heard the nesting pair's excited warning screams. Seeing me instead — a human, standing very still and not doing any harm — may have confused them; they dispersed almost as quickly as they had come. The nesting pair stayed. I still didn't move, and then one of them left as well.

The vocalizations of the pair at the nest with young had been aimed at me, as a warning to a perceived potential predator. The six other jays that had been recruited by their noise had provided no apparent help, except possibly communicating to the pair through their behavior that I was not a threat.

Do blue jays routinely come to another's nest if the nest owners sound an alarm? To make another test, I left the nest vicinity and came back a half-hour later to find out if the previous scenario would be repeated. Again only one of the pair was present, and this time it was not agitated. I was able to reach into the nest and touch the three almost black, featherless young without causing a ruckus. The parent that was present screamed only once or twice, shook its feathers once, and pecked at a branch, but soon appeared much calmer than before. I watched for another ten minutes, and after a while the jay seemed at ease and left. I left as well, planning to return later to deliberately induce the pair to again scream an alarm to see if the six others would make a second appearance.

When I approached, one of the parents was again on the nest. Again it flew off, and this time to provoke it I lifted one of the young out of the nest. This ploy worked: the parent confronted me up close and scolded me even more vigorously than before. The mate came instantly and gave short high-pitched, two-note calls plus rolling, liquid-sounding double cries with an undulation in each. Having no idea what they meant, I stayed and waited for further developments. The calls became louder, longer, higher-pitched, and more repetitive, and one of the pair flew so close to the back of my head that I felt a rush of air from the wing beats. It then showed its anger in front of me by hammering limbs and ripping leaves off branches. Since no harm could come from my presence as such (the myth that young birds will be abandoned if touched notwithstanding), I stayed for at least ten minutes, during which the pair's scolding never ceased. I heard no responses to the parents' alarm calls from other jays in the woods. I repeated the experiment/observation twice more on subsequent days and again eleven days after the first encounter, by which time the three young were feathered out, and each time I got the same negative result. Although these results don't fit the picture of jays' being attracted to others' alarm calls, they do not contradict it. The six could have been a group passing through, or if local birds they might have learned what the pair was excited about and lost interest in me.

Some species of jays have helpers at the nest (usually offspring from previous clutches) that assist a pair in protecting and rearing young — but none had been reported for blue jays. I now felt confident that there were no helpers defending this nest, and I was also sure the six jays that had been attracted to some of the pair's alarm calls had not come to help. The jays' investigation of alarm calls was information-gathering behavior: by finding out what had caused the alarm, they learned what had threatened other jays and thus should be watched closely or avoided. In contrast, the attraction of the mate and possible nest helpers to such calls is mobbing — ganging up on predators to chase them away. I suspect also that the so-called funerals attributed to jays that gather near a dead individual are a variation of, if not the same thing as, one of these responses.

My observations at the blue jay nest, and similar ones of a red-tailed hawk and a barred owl, convinced me that the jays' screams can act as powerful attractants, as the well-known lore suggests. But that does not explain why the thousands of other blue jay screams I've heard in the woods did *not* attract a crowd or even any individuals. Different meanings are attached to different calls, and a large part of the explanation could be variations in the calls — their volume, inflection, repetition frequency, pitch, and other nuances — that were beyond the scope of my methods and scale of investigation.

Blue jays also form another type of aggregation, one that is associated with excitement of a different kind. This excitement occurs only on the first warming days of early spring, from late February to late March, when blue jays get together, sometimes in large numbers, and spend hours in social palaver.

The few of these spring palavers that I have witnessed involved from eleven to more than thirty jays and lasted from several hours to most of a day. However, it may be that I usually saw these assemblies only at their peak, as I did at 9 a.m. on April 18, 2007, when twenty-five blue jays quickly gathered in the very tops of tall bare ash trees by a clearing on a hill. I watched in amazement as jays screamed, made rattling noises and bowing displays to one another, and flew around in small groups for a couple of hours. I had not seen much of the buildup to that blue jay convention. But one that occurred directly at my cabin on March 8, 2014, had a two-day lead-up, during which I was able to gather information about how it had started.

March 6 was one of those idyllic pre-spring Maine mornings, cold but clear. The sun was bright under a cobalt sky over a deep blanket of snow, and the trees were still bare. No jays were at the feeder at 7 a.m., so I was surprised to see six arrive in one group and land together in the top branches of a large spruce at the edge of my clearing, where they stayed silent and hidden in the thick branches. Strange, I thought. I waited a half-hour and neither saw nor heard anything. Could I have missed their leaving when I turned to put wood in the stove? I needed to know. Taking my axe to bang on the tree to flush them if they were there, I walked on the snow crust to the tree — and just as I reached it the birds flew out and scattered into the woods. Soon I heard jay talk spread well over a square kilometer, and the talkers included more than six birds.

Two days later temperatures had risen to 45°F, the warmest day of the spring so far. The sun was bright, and blue jays were suddenly in abundance near the cabin. One landed on the tiptop of the same spruce in which the six had perched silently, but this one called. Soon jays were sounding from the forest all around. Groups of two, three, or more began flying into my clearing and perching on the tops of the leafless maple trees. In twos and threes they bobbed up and down in what looked like pushup exercises, accompanying this display with a cacophony of high-pitched, double-note calls so close together that they almost sounded like one. They also made long rattling calls and at least six other distinctly different sounds. By 8 a.m. they were gone.

At 11:05 a.m. a group of eleven jays flew in, making a racket as during the earlier displays. Many of them were now in pairs, which perched together, performed vigorous pushups, and gave even more kinds of calls. By noon all had dispersed into the woods, where most of their calls reverted to the usual screams. Since jay palavers of this type occur only in the spring, a month or so before nesting, they probably function as mate marts, where single birds meet, display, and maybe size each other up and form pairs.

Blue jays also form large aggregations for reasons other than mobbing, bobbing, or mating. In April 2012 I saw seventeen flying together over the forest. And on May 19, when the pairs were busy at their nests, I sighted a larger flock than ever before: twenty-nine flying along silently in a ragged group. At that time of year, they were unlikely to be current-year breeders, nor could they have been the year's young.

Fifteen months later, on August 12, 2013, I chanced on a similar sight near my cabin. The sky was overcast, and it was late afternoon. I had not heard blue jays for weeks, but now a flock of about twenty-five flew over me. Like the other traveling groups, but unlike the jays at the spring conventions, they were silent. However, in the next weeks I often heard blue jays in the forest. At that time in late summer there were no acorns and no beechnuts, their favorite foods and/or staple diet in the fall. The first red oak acorns would within a week or so become available in isolated patches of forest. Maybe it was their time to go find them. Tree seeds and fruit are usually available in widely dispersed patches, and the birds feeding on them — finches, robins, and waxwings — travel in groups that provide many eyes and likely allow easier discovery of the patches, where the food's abundance reduces competition. The same principle may apply to blue jays' forming flocks for long-distance travel to find seasonally patchily distributed food.

Over a period of twenty-one years I eventually saw eighteen such flocks of silent birds flying low over the forest. Ten were in the spring and eight in the fall. I could not detect any pattern among them with respect to size, which ranged from five to thirty. The direction of travel was north/east in the spring and south/west in the fall. However, sightings were highly sporadic, and since I do not normally carry a compass, I gauged direction by eye. Also, since the birds flew low I may not have been able to see all of them; the numbers are minimums.

What moves out can also move in. By May 25, 2014, a crowd appeared to have arrived. After that day I heard jay calls from all around my cabin and saw jays flying, as singles and pairs, "everywhere." During an outbreak of small black beetles feeding on young tree foliage, I sometimes saw up to four in a tree at one time. A number of jays came to the birdfeeder, but given the calls in the surrounding woods, they were the minority. By May 30 there was still lots of jay activity in the vicinity. Jays were everywhere; I saw them in twos and in loose groups of several traveling together, and heard much chattering. From dawn until late afternoon their voices sounded from all sides of the clearing and far into the woods. The many different kinds of calls, with their various nuances, pitches, sequences, and frequencies, amazed me.

Blue jay numbers relate to the availability of food, which can be highly sporadic. In the fall of 2013 the beeches bore seeds, and on September 7 I saw jays after not seeing any in residence for months. Then on the afternoon of September 12 I heard a sudden and very localized clamor from a group of blue jays. I stalked into the woods to find the cause of the commotion, but could not discern a focal point such as an owl. Two jays would fly in one direction, three in another, while two more loitered near me; others were hopping around in a maple grove, and still more were scattered in spruces. Whenever some were together they made relatively soft calls. A half-hour later, as the jays dispersed broadly, I approached from another direction and encountered more excitement. Most of the jays were in dense forest where they could not have been in visual contact with each other. It seemed that their vocalizing might have simply been related to keeping themselves oriented to one another in a convention or social gathering. After that there were often several jays together at the beech grove nearby, but there, while gathering the nuts, they were almost always quiet.

These observations did not reveal the function of the group flights, only that jays could assemble and large groupings must have included more than one immediate family. It seemed the jays were not necessarily solitary at all, nor strictly restricted to a home territory. Yet even when their calling made the woods seem full of jays, they did not travel around in flocks within the woods. Are they nevertheless still a group, one that while traveling over the forest keeps in contact by sight, but when dispersed within the woods does so by voice?

Unfortunately I was unable to identify individual blue jays, except one that I nicknamed "the wheezer" (because of its distinctive nasal twang), a resident that I heard fairly often from September 12, 2013, until the next fall, and then again the following year. It is likely, though, and intriguing to speculate about the possibility that if the jays are members of groups they can identify many individuals by their aural signatures.

The rare and often dramatic instances of jays' coming together — to mob a predator, in a rendezvous for mate selection, or for long-distance travel — are overshadowed by their far more common tendency to be spread through the woods as individuals. But it raises the question: What does a jay gain by being solitary when it is alone in the woods? The answer may be found by determining what it does. I decided to follow one and find out.

November 18, 2011. It was a still and cold morning in the woods next to my cabin shortly after the sun came up. A blue jay near me perched high in a tall sugar maple tree, fluffed itself out, and swiveled its head to look here and there. I stood with my cup of hot coffee, greeting the sun and whatever else might come, expecting and soon seeing the pair of blue jays that had been daily for months to my birdfeeder.

I went back inside the cabin to warm myself at the wood fire in the big iron stove. Nothing unexpected happened until, some minutes later, a blue jay screamed the long high-pitched calls, one after the other and for several minutes on end. It was calling from the woods, about a hundred meters distant. Its persistence was unusual, and I wondered if it was alarmed or surprised by something, as per the generally accepted interpretation of screaming.

I peered into the woods to try to see some cause of the prolonged screaming. As usual the screaming bird was hopping around, seeming unconcerned, and randomly looking in various

A blue jay carrying an acorn.

directions. But then I heard a faint scream from the east, perhaps a kilometer away. I couldn't tell whether it was random or answering. Yet "my" nearby jay immediately stopped calling, and I again heard the call of the distant one. My jay called again, and the other replied, sounding nearer this time. Mine then stayed silent, but the other repeated its calls at intervals, and each time it called it was closer.

I had previously seen lone jays call from a treetop, fly off in silence, land in another treetop, call again, and repeat this pattern, thus moving in a specific direction. Now the distant jay was coming up the slope and mine flew to the eastern edge of the clearing, precisely where the other one soon arrived. They landed in the same (long since bare) red maple, but although seemingly ignoring each other made the soft whisper calls often made by couples. Then both came to my feeder, filled their throat pouches with seeds, and flew off separately, high over the forest. I realized then that a blue jay's home area can be huge. The forest is dense, and it may normally make more sense for a pair, or birds of a group, to forage separately than together, as long as they rely on scattered small food items that must be searched out one by one. But if these two were a pair, what about an extended family, or a social group?

November 6, 2013. It was a wind-still morning, and I was in the woods before 7 a.m. I soon heard the typical long jay calls, first to the north, then the south, east, and west. After several minutes of silence I heard another call, and then yet another from a half-kilometer or more distant. I sat on a rock and waited as a noisy pair of red-breasted nuthatches in a flock of chattering chickadees passed through the woods. Suddenly a blue jay arrived and landed about two hundred meters from me, high in a huge red spruce tree. I didn't budge. It was silent, hopping through the thickly branched top of the tree as though searching for food, as it picked here and there. A few minutes later, and without interrupting its exploration of the tangle of branches, it made a few of the usual long, loud, two-note scream calls. Sometimes it abbreviated these calls to make a series of three, varied the pitch, or slowed down or speeded up the tempo. All the while it kept hopping, pecking here and there, and foraging, without ceasing its calling.

At a distance of a kilometer or so I heard the faint call of another jay. The one I was watching paid no apparent attention to the sound, not interrupting its investigation of the twigs at its feet. In about twenty-five minutes it traveled only about three hundred meters. It made two or three different scream calls, one call with a higher pitch, and a series of lower, softer calls. I could see far into the leafless maple woods, and there was not another jay in sight. After a half-hour the solitary jay flew off silently to the northeast, from where I had heard no calling. On the face of it, it seemed to have talked to itself the whole time. But had it been heard and listened to by others?

These snapshots of a blue jay in the woods might not have been interesting except for the contrast with what happened two hours later in the same woods. I heard a red-tailed hawk scream and saw it sail low through a silhouette of trees, quickly out of my sight. (I was glad to see it, to confirm that it really was a red-tailed hawk, because the local jays mimic the calls of both the red-tailed and the very different-sounding broad-winged hawks to apparent perfection.) Almost immediately a blue jay started calling, then two, then three chimed in. Within a minute there were more jays and I could no longer distinguish individuals. Then there was a melee of calls, and yet another jay flew over me out of the distance and in the direction of the commotion. These were indeed the jay screams, but they had a different tenor from the lone foraging bird's previous calls that might have been heard but were usually not answered. Fifteen minutes later the jays became silent. Many jays must have been in the nearby woods before the hawk came, and likely afterward, and they would have heard the first jay, the one I watched, vocalizing as it foraged near me.

From then into early April I heard the jays' screams many times each day. Along a stretch of six miles (ten kilometers) of dirt roads I seemed to hear them everywhere. Groups of two or three flew here and there in no specific direction, and there was hardly a moment when I did not hear a jay calling from somewhere in the woods. But after April 7 the woods were again silent of jays. I saw none, as if the woods had been swept clear of them. Again only a pair remained by my cabin.

The observations are consistent with the inference that although blue jays are solitary in some of their food gathering in the northeastern forests, it is perhaps because they often have to be. Yet they broadcast specific loud screams into the otherwise silent woods, and these neither attract others nor warn others, but serve instead as distant markers of each other's or a social group's presence. Nevertheless, blue jays can and do recruit, but through other calls that also draw attention to themselves.

The scattered blue jays are sometimes in flocks, but they do not look like flocks while the birds are feeding separately on widely scattered food in a forest. On the other hand, movement to new foraging areas and clumped food such as the locally and seasonally available acorns and beechnuts require long-range movements, which are made by the group. Highly clumped nuts are available only in the fall, and the supply in a given location varies from year to year. In the fall of 2014, for example, unlike in 2013, there were no beechnuts at my study site in Maine, nor were there any acorns there. The acorn supply was spotty — in some areas along my route between Maine and Vermont the red oaks had few or none, and in others the trees were loaded with them. Food caching for future use is one way of dealing with a distribution of food that is uneven in space and time, but in the northern forests blue jays cannot rely on storing food for use in the winter, because the snow is often too deep for them to recover their caches. Having to move seasonally as well as geographically, jays somehow manage to maintain social cohesion and an apparent flock organization without staying in a flock continuously. They keep in touch vocally when separated, and can then assemble and move together.

I had now arrived at what I thought was the most reasonable hypothesis for these long, loud calls that were usually not recruitment signals, and had buttressed it with independent confirming evidence. Yet I wasn't satisfied. What could be the advantage of such seemingly costly behavior of "talking to themselves" (which might attract the attention of predators)? The reason that had come to mind, namely keeping in touch, just did not seem sufficient as a selective pressure.

"Animals," as Mark Twain authoritatively stated relative to blue jays, "talk to each other . . . but I suppose there are very few people who can understand them." The scream of the blue jay, I now suspect, has a basic message that simply says, "Here I am. How are you?" Hearers can then feel reassured that they are not alone in the area, and can either ignore it or reply, "Here I am, too." Depending on inflection, length, repetition, pitch, and context, it has subsidiary meanings such as "All ok," "I'm excited," or "This is scary, come check it out." Very different loud sounds uttered only at the spring rendezvous may mean "Look at me: I'm available." Soft whispers in the presence of a mate may mean "I like you. I want to stay with you."

In any relationship where the individuals are separated (such as by distance or by a visual barrier like a thick forest), giving a shout now and then — whether or not there is an immediate answer, so long as it is heard by the intended recipient — could be important without conveying a specific message. After realizing the likely nonspecific but nevertheless meaningful social nature of the blue jay scream, I thought that now I might be less hesitant to call, e-mail, or send a letter or a hello, even when, as almost always, I had nothing important to say. I had learned more from the blue jays by inferring something *from* them than by attributing a message *to* them.

10

Chickadees in Winter

BLACK-CAPPED CHICKADEES, ALONG WITH BLUE JAYS, ARE iconic birds of the winter woods. They travel through the forest in small flocks of usually fewer than a dozen. I may hear their cheeps and occasional *chick-a dee-dee-dee*s as they pass by while busily hopping among the branches from one tree to the next. The little fluff-balls in flocks grab attention by their cheerful energy and unflagging persistence in the coldest winters, and a day or two after the winter solstice the males start to sing their territorial or love songs.

I started watching chickadees closely and taking notes on their winter flocks, in which they were often in company with my favorite birds, the golden-crowned kinglets and red-breasted nuthatches. The flocks varied hugely in size and composition, and I hoped my observations would reveal something interesting, maybe some insight about the advantage of being in flocks.

If I had read Susan M. Smith's 1991 book *The Black-capped*

Chickadee: Behavioral Ecology and Natural History, which cites 610 scientific references, I might have realized that chickadees were many researchers' favorite subjects. Perhaps there was nothing more for me to see — or everything. I chose to believe in the latter possibility, simply because chickadees are ever present where I live. They are one of the few birds able and willing to stay year-round in their home in the north, enduring howling winds, snowstorms, sub-zero temperatures, and the constant challenge of finding something to eat in the barren woods. Birds of other species follow them not only as flock members, but also as onlookers. On July 21, 2006, a chickadee alarm-called to me, and almost immediately eight others, plus a junco, a red-eyed vireo, and a redstart, gathered round me. These were not all members of a flock, but the advantages of a flock had something to do with their behavior.

In the first week of January 2011, months after almost all insect-eating birds had gone south, I was cheered by two chickadee males' calling back and forth at dawn. It was still about four months before the chickadees would begin nesting. These two males were perhaps a hundred meters apart and were easy to distinguish because one gave his *dee-dah* call (sometimes described also as *fee-bee*) at a higher pitch than the other.

Nowadays the chickadees often seem to live on black sunflower seeds. A crowd of them at my birdfeeder in Vermont had consumed over a hundred pounds of the seeds per winter, and I could not imagine what they could have found as an alternative food source. But when I first began living at my cabin in Maine and put up a feeder full of black sunflower seeds, no birds visited it for weeks, although I met chickadee flocks almost every day I spent in the woods. I had no idea how many there were, or which ones lived and which died, or why.

Survival strategies of black-capped chickadees have long been of interest, and being interested in survival myself, I thought this a worthy topic. Aldo Leopold had sought answers to the puzzle of how they survived the winter, and in his classic 1949 book *A Sand County Almanac* he described the chickadees that came in the winter to a birdfeeder at his farm in Wisconsin. He captured and marked ninety-seven of them over a decade and learned from recaptures that only three remained in the fourth winter and only one survived to the fifth. By noting the farthest points from the feeder where his birds were seen, he learned that the chickadees' home range in winter was nearly a kilometer across. Leopold speculated that since the birds were well fed at his feeder, their most likely major killer was weather, and that the "genius" of the one that survived five winters was its ability to find shelter from the wind. However, he did not test the chickadees' dependence on wild food, and I wondered if food was also key to their winter survival. Might their habit of staying close together as a flock, besides reducing their chances of being killed by a sharp-shinned hawk, let them share knowledge about ways to broaden their diet to include novel foods as they become available from one season to the next?

Sunflower seed is a huge departure from chickadees' normal summer diet of insects, and given the variety and changing abundance of insects, there is scope for genius in keeping track of and finding them. I have on occasion collected cocoons of the Promethia moth (*Callosamia promethea*) in the winter that had been torn open. The contents of just one cocoon might be enough food to sustain a chickadee overnight at sub-zero temperatures. But to find one of these cocoons would be a major challenge, because they are wrapped in dead hanging leaves. Additionally, they are as tough as leather, and to open one would require a sustained effort, involving not only knowledge and skill but also confidence and persistence. These cocoons are rare, and those of all insects differ. It is a challenge to become aware of the rare, the new, all in the context of the profound change of summer to fall, and fall to winter, and winter back to summer. It seems to me that to stay abreast of the changing spectrum and availability of potential food in order to remain year-round in the same several acres may require constant exploring and an ability to adjust to unpredictable realities. In many locations survival skills now include finding a birdfeeder.

The feeder I put up at my cabin was at first an experiment; I wanted to find out if and when a chickadee would find it and feed there, and what would happen next. I was not confident that one would arrive soon, because chickadees remain in or near their home territory all their lives, and it was unlikely that any chickadee then alive in the nearby forest had ever encountered either a birdfeeder or a black sunflower seed. So what would induce any of these wild, un-suburbanized birds to visit an unfamiliar contraption, notice the little black objects in it, and hammer one open, thus discovering a new food source? And if one bird did clear all those hurdles and discover this food, would others come to eat it also? Might they learn from one another, resulting in a sudden influx of sunflower-seed eaters?

I hung my birdfeeder loaded with sunflower seed from a branch of the white birch by the cabin door one December. I waited a month. No birds came. Maybe the location was wrong. I moved it into the woods about three hundred meters away from the cabin. Two weeks later, it was being visited by both chickadees and blue jays. I transferred it about fifty meters in the direction of the cabin. Chickadees came to the old location, searched, and one found the new site within minutes and was immediately joined by others. I then moved the feeder another fifty meters, and they almost immediately found it there also. I had the impression that as soon as one bird found it, the others at the old position flew to the new. In less than a half-hour of several more steps they were all visiting the feeder now hanging on the white birch tree next to the cabin although at the original site they had ignored it for a month. I had hoped that the chickadees might tell me something about how they survive in the forest through the winter, and also about how they thrive even as their world changes from month to month. The birdfeeder experiment suggested that being in flocks might be part of their solution, because the birds were sharing their discoveries by learning from one another.

The chickadees at my feeder in January 2015 appeared to be a flock of up to sixteen individuals, the maximum I ever saw there together. But another experiment showed that there were many more around. Over five days I made 131 counts of their numbers at the feeder, for a total of 708 chickadee sightings. I could identify two individuals — a partial albino with a white cap, and a bird with a thin strip of orange flagging tape on its leg — which together accounted for 15 of the 708 sightings. Thus, assuming that these two were representative of the rest of the flock, there were about ninety-four chickadees in total.

Chickadees usually forage in loose, species-mixed flocks of from two or three to over a dozen individuals. As a routine exercise in my walks and stays in the woods over the last twenty years I had collected data on winter flock size, composition, and behavior. Since 1980 when I started deer hunting in the fall and teaching winter ecology in Maine, I have conducted most of this late-season birdwatching while sitting in a tree, usually the same one, a large balsam fir. Over the last thirty-four years I have spent 1,700 or more hours in that tree, opportunistically watching chickadees plus the other birds that sometimes accompanied them: several golden-crowned kinglets, a brown creeper or two, a pair of red-breasted nuthatches, and occasionally a single downy woodpecker (but never any other kind of woodpecker). I didn't and don't know what brought the various species together or who was attracted to whom and when, why, and for how long. But I did discover a good reason why chickadees are at times found closely associated with red-breasted nuthatches.

Red-breasted nuthatches live in conifers, and in the winter they feed on the trees' seeds, which they pull out of the cones hanging from the topmost twigs. In Maine, a variety of conifers — red and

Chickadee on a cedar bough, with microlepidopteran moth caterpillars inside the evergreen leaves.

white spruce, white pine, hemlock, and tamarack — have seeds available in winter, though not every winter. For reasons that are not understood but may involve subtle weather cues, trees of any one species tend to flower (and later fruit) synchronously, but not all species produce seeds at the same time. Chickadees are associated most conspicuously with red-breasted nuthatches in years of huge winter seed crops. A reason for this association is not hard to guess: when the nuthatches pick at the cones in the treetops they cause seeds to spill, and the chickadees forage for the seeds scattered on the snow below. It's an example not only of having a broad diet but also of the same alertness that allows chickadees to find and adapt to a birdfeeder stocked with seed.

The winter of 2010–2011 was unusual in my part of Maine in that no trees of any kind were in seed. It was thus a natural experiment. One of the first, perhaps predictable, results was that the woods were empty of red-breasted nuthatches, which had presumably migrated to look for food elsewhere. Another result, perhaps not so predictable, was that while the chickadees were still there, they appeared in smaller flocks. I saw flocks in October, but by November most chickadees I encountered were alone or in groups of only two or three. Had they previously gathered in groups near the nuthatches simply because they were attracted by the dropped food? In addition, what were the chickadees feeding on now?

I already knew that chickadees are quick learners and flexible in behavior. They build their nests mostly in the deep woods in cavities they hammer out of rotting wood. Most of the cavities I had seen them make were in dead birch stubs. It is a mystery to me how they penetrate the solid leathery bark of these trees, as if knowing that under the thick bark they may find soft, rotting wood. How can they know, given that a live tree's outer bark is identical to that of a dead one but the living wood underneath is far too solid for them to excavate? Since they start holes and then leave them, may hack out a nest cavity fifteen meters up in a sugar maple or poplar stub, and accept almost any kind of bird box, they are likely experimenters able to adjust to a wide range of options. Whatever cavity they use, inside it they build a deep and cozy nest — of moss, animal hair, plant fiber (such as the fuzz on opening fiddlehead ferns), and down feathers — to harbor their clutch of six to eight eggs and young.

Finding their nests is always fun. The challenge is the unpredictability. My latest discovery was especially so. As I walked up the path toward my cabin, I saw a chickadee with a green caterpillar in its bill and stopped to watch it, hoping to see it slip into a nest hole. But it stayed near me and kept flitting from branch to branch, so I knew the nest was close. I spotted a likely site, a dead gray birch stub, investigated it closely, and was surprised not to find the expected nest entrance. Seeing no other potential nest site nearby, I drew back to gain more distance from the bird — and saw it fly to the birch stub I had just investigated. I had been fooled: the nest entrance these birds had hammered out was at ground level. My surprise at their choice made me smile, and the memory was precious, so I made a drawing of it. (The next year, in 2015, a pair nested in a less picturesque location, a bird box in the open on a post alongside the garden. Red ants raided the nest the day the young hatched, but three days later the parents were nesting in another box near it.)

How the chickadee had found this and presumably many other little green caterpillars to feed its babies nestled at the foot of the birch stump was not the result of a simple hard-wired response. Caterpillars such as those prized by birds have a bag of tricks that make them hard to locate. They roll themselves up in leaves, or camouflage themselves by aligning themselves along the leaf centers so they look like the midrib, or mimic the leaf edges, to which they attach themselves to fill in for the part of the leaf they remove. Sometimes the main clue to a caterpillar's presence is the feeding damage it causes on a leaf. And even then, the caterpillars that are hunted by birds have evolved tricks that help foil their predators. After feeding they distance themselves from the evidence of that activity left on leaves, or they clip off the leaf remnant and then hide while digesting their food. Still, feed they must, and leaving some evidence is unavoidable. I wondered what clues chickadees used to locate the almost invisible prey. And in the summer of 1981, in the woods next to my cabin in Maine, a colleague and I discovered the answer by experiment.

My fellow biologist Scott Collins and I captured chickadees and housed them in a screened enclosure that we built in the woods. It had two separate parts, each 2.5 meters high and covering an area of 30 square meters. We placed tarps on the bottom of both compartments to discourage the birds from trying to find food on the ground. One compartment, used as the holding area, enclosed naturally growing small conifers and other perches and had a closeable flap to and from the other compartment, which was used as the training and testing area. We held confined only one bird at a time (and released it immediately after a series of tests), except for one mated pair along with the birch stump containing their nest with five nearly grown young. The other birds in the experiments, which we knew were all males because they sang, we individually identified as Ralph, Frank, Fernald, and Duke.

Each day we cut and placed in the training/testing compartment twenty leafy trees (chokecherry or birch) one to two meters high. In two of the trees (placed at a different location each day) we made holes in some of their 700–1,100 leaves with a paper punch, or (in some tests) the leaves had naturally occurring leaf damage made by feeding caterpillars. For the training procedure we tied segments of rose twigs onto the experimental trees (those with holes) and impaled half a mealworm on each thorn. For the tests we removed all remaining bits of mealworm to find out if and when the birds might use leaf damage as a hunting cue.

The chickadees immediately examined leaves closely when we released them into the training/testing arena, but we found large individual differences in their behavior. The mated pair and Duke showed highly significant preference for hopping to and foraging on the trees with damaged leaves on the very first trial, before having received any rewards from our training, suggesting that they had already self-trained in the wild. Fernald, who did little inspection of individual leaves, usually flew directly to the tree with the damaged leaves when let into the foraging arena. Ralph, on the other hand, who at the beginning showed a preference for foraging on the ground and a slight avoidance of the trees with

A chickadee observing leaf damage.

damaged leaves, after ten learning trials switched to show a highly significant preference for trees with damaged leaves.

Since our chickadees learned to hunt the caterpillars by favoring trees with fake caterpillar feeding tracks, and subsequently were also able to differentiate between different kinds of trees, different kinds of caterpillars, and even different kinds of leaf damage, perhaps they could also learn about novel sources of food from one another.

Learning involves the ability to remember, a proven faculty for chickadees, which is useful in the winter when they store food in caches for later retrieval. And this faculty is not irrelevant to humans. Fernando Nottebohm of Rockefeller University and his co-workers discovered that when birds learn to sing, and also when adult chickadees in the wild start to store seeds and exercise memory to retrieve them, their growth of brain cells accelerates, and when they stop those activities, nerve cell deaths follow. We had been taught that we humans start losing brain cells in our twenties

European starling

Yellow-bellied sapsucker pair

Young sapsucker (painted after it was killed at a window)

Barred owl

Broad-winged hawk

Blue-headed vireo

Red-breasted nuthatches

Blue jay on
American chestnut

Chickadees at nest site

Ruffed grouse feeding on
hornbeam flower buds

Great-crested flycatchers

Male red-winged blackbirds in transit

Red-winged blackbirds in marsh

Evening grosbeaks feeding on young poplar leaves

Woodcock

Baby woodcocks in hiding

and keep losing them all our lives. But these findings about birds suggest that exercising our minds may create more brain cells at any time in life. (I'm not sure, though, if researchers who study humans make a distinction between those who maintain a chickadee lifestyle and those who don't.)

Following chickadees in the winter over the years, when the trees had shed their leaves, I often wondered what they fed on after they ran out of stored sunflower seeds, or memory storage space. It seemed there was no obvious focus on specific objects in their foraging; they were pretty much everywhere examining pretty much everything, just as our tested chickadees did when we released them from the holding area into the compartment with twenty trees with potential food. While following a chickadee flock in the forest in the winter, I usually saw individuals alternately visiting the bare tops of tall maples; the crowns of pines, fir, and spruce; and the vegetation close to or even on the ground. They picked at lichens, dead leaves, little twigs, bark, large branches, and tree trunks of almost any species. I suspected that a large part of their foraging involved continuous exploration of everything so

From left to right: *Chickadee egg, just-hatched young, and half-grown young.*

they might identify any potential food. They did not specialize in any specific kind or part of the tree.

But then I encountered a glaring exception that would prove the rule. Late in the winter of 2010–2011 I began to see flocks of chickadees foraging daily in a white cedar swamp, where I had not seen them linger before (or since). What were they finding near the tops of the cedar trees day after day?

The snow under the cedars was strewn with bits of dead twig tips, and it was easy to see that the chickadees were picking at such tips in the trees. I followed one flock of seven as it traveled to, and worked on, one cedar after another. These birds eventually left the cedars and went to the tops of red maple trees, where they continued to pick and dislodge debris onto the snow, then left the maples and went to the understory of some balsam fir trees. All the while, they stayed together in a loose flock.

Wondering what attracted the birds to the cedar tips, I took some cedar branches back to the cabin for a close look. Examining the discolored cedar tips under the microscope, I found minute caterpillars inside almost every browned leaf. The small leaf-mining caterpillars were those of a tiny gray moth, *Argyresthia thuiella* (of the microlepidopteran moth family Yponomeutidae). They overwinter at the tips of branches inside the scalelike leaves of the eastern white cedar, killing the twig tip and making it prone to dislodge from the tree when birds peck at it. They pupate in spring and the moths emerge in June. Might the chickadees have been feeding on them? Given that chickadees in other years had concentrated much of their foraging effort on nuthatch-spilled conifer seed, it seemed unlikely that they were now eating frozen caterpillars (recent temperatures had dipped to −25°C). And I had no way to see what the birds high in the cedar trees were ingesting. But later that month, thanks to a northern shrike and an accident, I found out.

The shrike had visited my clearing when a group of chickadees was near the birdfeeder. I noticed it only after seeing the chickadees either panicked and fleeing or perched frozen in place on a twig. One hit a window and was killed at once. Not wanting to waste a bird, I skinned it, noting that it was fully muscled and had small amounts of fat under the skin near the throat and tail areas. Instead of sunflower seeds, its bean-sized gizzard held a mass of stringy matter mixed with dark bits and pieces that could have been of either plant or animal origin. I could not tell what this material was, but when I spread out the jumble of bits and pieces in alcohol in a petri dish under the microscope and rummaged around in the partially digested "stew" for a half-hour, I began to have an inkling of what I was looking at.

The first clue came from what looked like an insect spiracle (an external respiratory opening) embedded in transparent skin-like tissue. A detached jointed leg confirmed that it was from an arthropod. Perhaps the bird had eaten spiders. Then I found unmistakable proof of a caterpillar: a head capsule. Eventually I also found a caterpillar leg attached to something long, fibrous, and transparent that had the spiracle and looked like a collapsed plastic bag. Eureka! I knew then that they were caterpillar exoskeletons emptied of their contents. There were many of them: I counted twenty-three. I also found other tiny, hard, serrated objects of at least three different kinds, then one of them attached to a caterpillar head capsule. Eureka again — they were mandibles of very tiny caterpillars.

Caterpillar mandibles, like those of other insects that chew hard material, have serrated edges that are analogous to our teeth. The dentition pattern is specific to a species. I found three kinds of it, so the bird had consumed at least three species of caterpillars. I immediately reexamined the cedar tip caterpillars: their mandibles matched one of the three kinds I had found in the bird's gizzard. Here was near proof that the chickadees were feeding on at least three species of caterpillars, one of which was the microlepi-

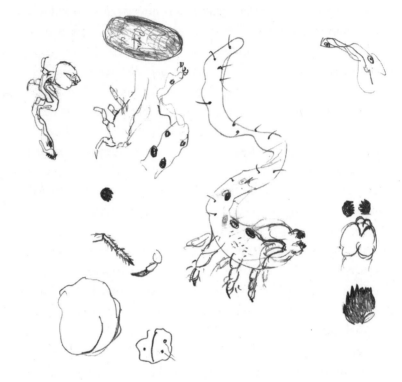

Remnants of one chickadee's stomach contents, showing caterpillar hard parts.

dopteran caterpillars from the cedar leaves. There was only one local source for that species: the trees in the white cedar swamp where I had seen the chickadees routinely.

It seemed possible that a chickadee could have happened upon these caterpillars. But flocks of them? I was now left with another question: How had the many chickadees discovered this food hidden within the cedar leaves? How had they learned to associate the brown leaf tips with caterpillars that were not visible?

I wondered if chickadees automatically examined anything that they had not seen before. To find out, I did a quick experiment with some chickadees that had been coming to the suet I had put out near the cabin door. After they had come to feed for two days I removed the food and provided instead a variety of items, including a red coffee mug, a beer bottle, a red-handled corkscrew, a marshmallow, a piece of wax, a matchbook, some cooked squash, a piece of a polypore fungus, and some coffee grounds. To my surprise the chickadees did not approach the visually most striking objects. Instead, a couple of them hopped close to and briefly examined a pile of coffee grounds, a marshmallow, and a squash rind. But the items that drew them immediately were some grease I poured onto the snow, a piece of red meat, and a bacon rind hanging from a twig.

Three chickadees that had been coming to the suet were hopping and flying about in a perpetual motion, as is typical of their species. However, on one occasion while I was watching them from the window they stopped, frozen in place, and stayed so for several minutes. I suspected that a hawk had come near, but instead it was the aforementioned great northern shrike. It was perched on the top of a maple tree about two hundred meters from the cabin. At that distance I at first mistook it for a blue jay: jays and chickadees routinely come to the same feeder and ignore each other even when relatively close together. The chickadees obviously had not made that mistake; shrikes routinely hunt songbirds. I seldom see more than one shrike in any one winter. I therefore suspected that it was not long experience that enabled each chickadee to recognize a shrike from a long distance and distinguish it from a blue jay, and that it was unlikely that all of them would have spotted and recognized it nearly at once. The chickadees' reaction must have been in part a social response. And if chickadees could learn about a dangerous enemy from one another, why not also the often quite specific appropriate food? It seemed credible that a chickadee or two had discovered the caterpillars within the leaf tissue of the cedar leaves and that new knowledge had spread to the rest of at least one flock.

. . .

It was not until three years later, on December 15, 2014, that I saw the chickadees' reliance on each other in action. Next to my south window a patch of vegetation was sticking up out of the snow. There was a spruce tree with a young pin cherry coming up beside it. There were dried stems of fireweed, grass, and goldenrod. And directly in front of my desk through the window, I saw a chickadee pecking, pulling, and tearing into something held within a dry curly brown leaf. If not for this chickadee, I would never have noticed that leaf, because it looked to me just like the others attached to the goldenrod and fireweed stalks. The other chickadees, over ten meters away at the feeder, seemed not to have identified it as of interest either. But quickly one came within a meter of the first bird, hopped around near it, and in a few seconds was joined by four others. I had, in weeks of watching, never seen even one at that location. The birds' heads were swiveling in all directions and they were picking at asymmetric shapes on leaves and twigs. For a time they left the first chickadee alone as it continued to peck at the debris, but after pecking here and there and finding nothing, they came over to the one that *had* found something. I too wanted to know what it had found, and rushed out as soon as the bird had left, and picked up a now totally empty spider egg case. The others could not have come near this bird because they saw *food* — they had come because they saw the other's *behavior:* acting as though it had found food.

Two days later I again was at the window. I had replaced the remnant of the spider egg case, and there were no bird tracks on the snow next to it. But now I noticed a similar silk package on the same vegetation-clad bank — because I saw a chickadee fly to it. This bird spent only three or four seconds pecking at it and then flew on, but a second bird immediately took its place and stayed a minute, and then another, another, and still another. The first bird that had pecked at it had not managed to penetrate it and had found no food. After the second bird departed I went outside to gather the object of interest, which turned out to be an egg case of

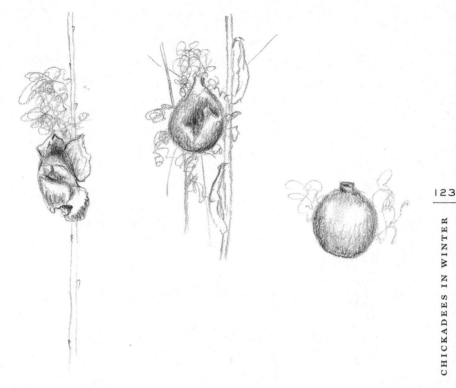

Left and center: *A garden spider egg case opened by a chickadee to gain access to the eggs.* Right: *An intact egg case.*

a garden spider, *Argiope aurantia.* There were only a few tatters on it; it was not penetrated.

A month later I again saw a lone chickadee in that tangle. It appeared to be searching and stopped to pick at some curled dry leaves. Another chickadee joined it, picking at the empty curled leaf the first one had inspected, and seconds later a third chickadee did the same.

Chickadee see, chickadee do. Evidently one advantage of chickadees associating in flocks is making discoveries and pooling what they see and what they learn about it. Society does the same for us. It provides us with many opportunities that may come our way without our seeing them on our own. But sometimes we do see them — and seize them — and that makes all the difference.

11

———❧———

Redpolls Tunneling in Snow

NATURE IS A MAGICIAN WITH A BIG BAG OF TRICKS. SHE pulls rabbits out of a hat, and it is a joy to see what turns up. Sometimes what you get is different from what you expect, and that is when the magic starts to come alive.

Chickadees and their kin may stay all their lives in the same small flock in the same area, where the individuals probably know each other and have a dominance hierarchy. Northern finches, in contrast, often form huge anonymous flocks, and they specialize in and hunt for specific food over enormous distances. Their food tends to be seeds that are packaged by the plant's fruits in various ways that make them mechanically difficult to access. Some grosbeaks have massive thick bills that crack cherry pits. Goldfinches have delicately pointed bills to handle the small seeds of thistle heads. With their specialized bills, crossbills pry the bracts of conifer cones apart to reach the seeds at their bases. But all these finches have one thing in common: the need to be at the right place at the right time to keep up with plants' unpredictable fruiting schedules. The seed eaters move, as I've said, sometimes over vast distances, and so become unpredictable as well (at least to us), often being rare or absent for years in a geographical area before becoming abundant. In North America these birds include two species of crossbills, pine grosbeaks, evening grosbeaks, pine siskins, and redpolls.

The common redpoll, *Corduelis flammea,* is one of the smallest finches at home in the high Arctic snow-swept tundra. It lives year-round north of Hudson Bay, where it is a seed specialist on the cones of spruce, birch, and alder. Redpolls come south in winters when their food crops fail in the north. The white-winged crossbills had most recently come to Maine in the winter of 2006–2007 and the pine siskins in 2011–2012. And a large movement of common redpolls into the northern United States occurred in the winter of 2012–2013.

In 2012 the first snow fell at my cabin in Maine on November 5, and eleven days later a flock of about thirty common redpolls appeared at my feeders filled with black sunflower seeds. They had not ventured this far south from their high Arctic domain for many years, so it was a treat to see them. They continued to arrive in ever-greater numbers until by January 20, 2013, a flock of about a hundred, and sometimes as many as 150, came to my clearing and my feeder daily. Throughout this time the snow was fluffy and temperatures ranged from −8°C to −24°C. A flock of up to twenty-two evening grosbeaks, twenty pine grosbeaks, and eight purple finches were also regular visitors. So it was not surprising when on January 23, beginning four minutes after sunrise, approximately 150 redpolls as well as evening and pine grosbeaks and purple finches arrived again. Working at my desk at the window, I at first paid them scant attention, because I had by then seen them often. The wind was blowing and the temperature was −26°C.

As usual after their early-morning feeding, the redpolls started hopping around on the deep powdery snow. They concentrated their activity in and near a patch of chokecherry bushes, even though the feeders were still loaded with food and seed had spilled onto the snow. Although no food was visible where they were gathering, I noticed one bird after another duck headfirst into the snow and then, while fluttering its wings, push itself several centimeters forward with its head still submerged in the snow.

The redpolls' odd behavior looked somewhat like typical birds' bathing, during which the head and front of the body are ducked down and quickly brought up and the wings then beat and spray water. However, the redpolls' behavior in the snow differed. It included simultaneously pushing and walking forward, and it lacked the repeated head lifting but had the wing beating. After each burrowing episode, the redpoll popped up and resumed hopping on top of the snow, having left a groove or a tunnel in it. I had not seen this before, so I took the opportunity to gather details.

By the end of that day, fifty-one of these groovelike depressions were scattered about where the numerous individuals had gathered in small groups. Most of these artifacts of their snow bathing were 4–5 cm wide, 6–20 cm long, and usually 4–6 cm deep. Five were tunnels, having clear entrance and exit holes with an intact bridge of snow between them. The birds' hopping along on top of the snow had left only tracks that barely indented the surface.

After approximately an hour of this activity the birds returned to the feeders or perched almost motionless in small groups in trees at the edge of the clearing. By early afternoon they had all left the clearing. No bird had stayed in its groove or tunnel for more than a few seconds.

I had no idea what to make of the behavior (although there were some similarities to the winter behavior of tunneling ruffed grouse; see Chapter 12). There had to be a meaning, a story, behind it. Were the redpolls bathing, looking for food, or making shelters to protect them from the cold? It was a puzzle, providing an impetus to look closer.

Redpolls hopping on snow, and one burrowing in it.

As I have mentioned, in this winter of 2012–2013 the common redpolls were not alone at my feeders. Four additional species of northern finches and four species of other birds were also present for weeks at a time. So this was an excellent opportunity to compare the behavior of redpolls with that of other birds, especially the concurrent finches.

The next day, January 24, when the temperatures had dropped to –28°C during the night, about a hundred redpolls showed up after first light. Again, soon after feeding on sunflower seeds they did some snow bathing, but a strong wind obliterated the record of their ground activity so I could not get data. The cold and snow lingered, and on January 25, at –23°C, after the dawn feeding the redpolls gathered on the snow in the same chokecherry bushes. This time they left a total of fifty-nine snow furrows. A group of them gathered in another patch of chokecherries, where they left fifty-four more furrows. A third flock gathered in still another small clump of chokecherries and left sixteen furrows. Seventeen more furrows were also left in the open at the periphery of the clearing, but only three furrows were made within the large open areas of the field itself. As during the two previous days, the burrowing/snow bathing occurred only after the birds' morning feeding. It was over by 10 a.m., but small groups of the birds continued to come and feed (but not burrow) until about 2 p.m.

Low temperature alone could not have sparked the behavior, because on January 26 the weather was scarcely changed — it was still as cold and clear with no wind and the snow was as before — but there was no snow burrowing. And on January 27, with similar weather conditions, only forty-two new depressions/tunnels were made. That was the last day I witnessed snow bathing, although the redpolls and other birds continued to come daily all winter. The weather had generally been cold with dry snow on the ground, and for snow bathing the redpolls would have needed wet snow.

The redpolls commonly roosted high in the trees around the

clearing, never staying more than a few seconds in their snow depressions. They went to the forest at night but the snow surface there remained free of their tracks and anything resembling furrows or burrows. On the other hand, in the forest I routinely saw the tracks and overnighting snow caves of the much less common ruffed grouse, so it is likely that the redpolls overnighted in the trees.

The redpolls' behavior increasingly seemed odd to me because during my five months of observation, the grosbeaks, American goldfinches, and purple finches had done no snow bathing or burrowing. I went to the literature to search for previous reports of snow tunneling and suggestions about its causes. In the winter of 2001–2002, flocks of redpolls had arrived in Adirondack Park in mid-January and been seen fluttering in the snow. J. E. Collins and J. M. C. Peterson concluded in 2003 that the birds stimulated one another in this activity and were apparently doing it "for enjoyment." I presume that enjoyment is a given — eating is for enjoyment too, but we know its function. R. S. Palmer (1949) and G. Furness (1987) had described similar bathing activity of common redpolls in Maine. In contrast, small birds in northern Europe, including redpolls and Siberian tits (*Parus cinctus*), had been reported to take up residence in snow burrows to escape wind and cold. To try to test these hypotheses, I examined the data on the redpolls as a function of the weather. Temperatures varied from –28°C to 4.5°C, skies from clear to overcast, snow from powdery to sticky wet. The redpolls needed powdery snow to tunnel in. But they did not bury themselves in the tunnels they made; moreover, they never bathed in wet snow, and they tunneled only in cold conditions but did not overnight in the tunnels even in the coldest weather. With hundreds of redpolls present around the feeders for many days over a wide range of weather conditions and temperatures, no factor of the external environment stood out as an obvious trigger for the snow tunneling.

However, individuals within the redpoll flocks are strongly influenced by one another. They arrived in tight flocks as in a small cloud, remained in tight flocks when they were feeding, and left the same way. Flight stimulated flight. When one flew to feed at a spot where I had not seen any birds in weeks, another and another quickly joined it until dozens were assembled there. In one instance approximately twenty redpolls gathered where one bird had first flown to pick at the seed head of a pigweed (*Chenopodium*) exposed above the snow. The twenty returned to this site at least fifteen times in succession in short, in-unison flights, but none made an attempt to burrow into the snow where the seeds of the mostly still-buried plant could have been found. Similarly, bathers attracted each other and also acted in unison.

The tunneling behavior could not be explained as a foraging strategy, since it did not occur in the presence of food but rather in its absence after feeding was finished. The default hypothesis is that the making of furrows and burrows in the snow is related to building shelters for energy economy and warmth. As we'll see in the next chapter, the local grouse submerge themselves in loose snow and make a new snow cavity each night. Small body size should make the redpolls much more vulnerable to cold than the grouse because it accelerates heat loss, and they may compensate behaviorally by huddling, as golden-crowned kinglets do on tree branches at dusk. Redpolls, though, don't huddle, and likely have other ways to survive at sub-zero temperatures.

One important consideration is that these redpolls in Maine were far to the south of their usual range. How they survive winter nights in the Arctic tundra is not known, although some researchers have suggested that they overnight in snow tunnels. But since even in wind and at temperatures to −26°C they remained in their local burrows no longer than several seconds and did not return to them in the evening, this explanation for their tunneling in Maine seems questionable.

Here in New England, the burrowing serves no immediate function. The usual reason given for apparently senseless behavior is play. Though motivated by enjoyment rather than anticipated reward, play provides practice for adaptive behavior that may be needed in the future. For example, redpolls obviously have not only the ability but also a strong urge to burrow in snow. This behavior may be fully expressed in the Arctic, where it is useful to help them survive the cold, and only partially expressed, as play, in places where remaining in the burrow may be costly. For instance, small birds in burrows under snow in New England forests would be vulnerable to ground predators such as short-tailed shrews and weasels.

But the main reason for not overnighting in their snow tunnels in New England, I suspect, has to do with temperature fluctuations. During most winters here, days of snowfall, snowmelt, and deep frost alternate, often overnight. A small bird in snow that is alternately subjected to thawing and freezing is exposed to the lethal risk of becoming trapped under an ice crust. Frequency is of little relevance, when just one such ice-storm event could kill an entire population if all the birds made the mistake of burying themselves in wet snow the evening before an overnight frost. Perhaps the redpolls truncate their overnighting behavioral repertoire in the south because of the more variable winter temperatures.

Redpolls were back two years later. They were in smaller numbers, up to twenty at a time. As before, temperatures were low and the snow was powdery for the two months or more that they regularly visited my birdfeeders. They hopped on the snow daily, but I saw none of the bathing/tunneling behavior. I was and am hugely pleased to have used the rare opportunity to observe their snow bathing in the brief period two years earlier. The absence of the behavior since then makes it all the more intriguing, and shows that there must be more to it than I imagined.

12

Tracking Grouse in Winter

THERE ARE RUFFED GROUSE IN MY WOODS, AS IN ALMOST all woods in northern North America. I love the sound of the males' music on warm spring mornings, and I have also heard that deep, pulsing, thunderlike drumming in the night. I see them most often in November, when I'm likely to be up in a tree in the woods at dusk waiting for deer. It is quiet then and I'm straining my ears hoping to hear a twig snap from perhaps a deer's distant footstep. At such times I have often been startled by grouse, because beginning in the fall they venture off the ground and fly into the treetops for their brief evening feasts on buds. At dusk in the otherwise silent forest you hear every stray wing beat against a branch, and the sound is startling, especially in deer season after you have been hyper-alert for hours at a stretch, day after day.

Grouse are our most hunted northern game bird, and they are the favorite prey of goshawk, red-tailed hawk, and great horned owl. Perhaps for these reasons, they are one of the shyest of birds.

But rare individuals at times apparently become attached to humans. A friend of mine, a logger, told me of a grouse that would meet him every day in the woods and follow right behind him in his skidder. Another friend, a naturalist, had one follow him on his bicycle (he stopped and grabbed it for catch-and-release). A grouse visited another friend's family on their porch every day for months. A photo in *Northern Woodlands* magazine (winter 2013) shows the naturalist Mary Holland with a wild grouse on her knee — the bird had followed her on her hike through the woods. She later reconnected to it on several occasions. From me grouse have mostly flown away, and pronto, except for some females that seemed intent on attacking me when I inadvertently approached their young. On the other hand, grouse do not flush easily from a nest when incubating the eggs. Only when I've come near one repeatedly has the female flown off and out of sight before I got there.

Ruffed grouse stay year-round in the north woods no matter how severe the winter. I knew something about their lives in summer and fall, but nothing of them in winter except that they seek overnight shelter from the cold by burying themselves under the snow. Snow insulates, so this behavioral adaptation to winter reduces their need to shiver to keep warm and hence conserves energy. Grouse in the winter woods of Maine face no shortage of tree buds, but energy extraction from this mostly roughage diet is limited by the processing capacity of the gut. As with most herbivores, a grouse's gut has a full-time job just to keep up a positive energy balance.

By January 2015, as in many Maine winters, we had not only an abundance of snow but also nights and days of sub-zero temperatures. Curiously, although it had been a good year for grouse, I had not seen one for nearly six weeks. But then, while I was out on snowshoes looking for woodpecker holes as possible overnight shelters of chickadees and nuthatches, a grouse exploded out of the snow within two meters of me. This would hardly have given me pause except that I knew grouse will spend the *night* under the snow. Given the context of the birds' and perhaps my own problems in overnighting at sub-zero temperatures, I was puzzled. It was, I noted at the time, a "sunny and practically warm day, and the sun had been up for four hours — so why was this bird just leaving its bed?" Flushing a grouse out of the snow at noon was not unheard of — I had done it in earlier years — and it was easily dismissed as common knowledge, at least to some. But common knowledge is often worth examining.

I had dug into grouse snow dens in the past, and given the birds' diet, had not been surprised to find numerous scat. Grouse scat is nothing like the white liquid splotches we usually associate with bird droppings. A diet of tree buds, often those of flower catkins, creates firm, sausage-shaped pellets that look and almost feel like catkins themselves. I examined this den, too, and found fewer scat than I expected. This could be significant, because with the bird sleeping late there should have been more rather than less scat. I counted only twenty-two scat. I didn't know how many the grouse normally produced overnight, but I did know I had a question — and a reason for counting the scat of many more of their snow dens.

There had been a series of light snowstorms on top of a couple of heavy ones; conditions were ideal for grouse to den in the snow, and the depressions that marked them on the otherwise smooth white snow blanket could be seen from over a hundred meters away. I immediately started taking more and longer hikes through the woods.

In the course of two days I found a couple dozen previously occupied grouse dens and collected the scat piles, put them into plastic bags, and thawed them at the cabin to count the pellets. The number of scat in a den varied from three to seventy-one, although most dens contained forty-five to fifty-five. These data and other observations associated with the dens suggested various possibilities that I had not thought of before, and so I was eager for more samples and more observations. Unfortunately, even after collecting many more scat samples I still could not make sense of it all, because most of them were from old dens that had been formed in the previous weeks when the light snows had only partially obliterated this track. What I really wanted to know was the time of day when each grouse had entered its den and when it had left. But could I find out?

I knew when a grouse had left its den only if I was lucky enough to flush one out of its resting place. But I would not know when the bird had entered its den, or how long it would have stayed if left undisturbed. Theoretically, I could have learned entrance and exit times by staying hidden in the woods and waiting and watching, for maybe a month, maybe more, until a grouse happened to come by and make a den within sight. If I then continued to watch until the grouse emerged from the snow, I would know the precise length of time spent in the den — by one grouse, at one time. Clearly, that strategy was not practical.

An alternative possibility — the one that I tried — involved making, and then often traveling, a long snowshoe track through the woods. On twice-daily runs I recorded the presence and progress of dens as they revealed themselves from the disturbance in the snow that marked a den entrance and the equally conspicuous exit hole. I took into account the time of day, the timing and force of occasional winds that disturbed the snow, and the intensity of snowfalls, all of which information would help me estimate how long a given den was occupied. During each run I took the scat for counting. After all previous dens had been accounted for, any entrance or exit holes on the next run would be new. Given that my view extended at least twenty meters from the sides of my path, a two-kilometer loop covered a considerable area. I used my data on den residency times along with scat counts as a roughly calibrated scat clock from which the length of time a grouse stayed in a den could be estimated by the number of scat.

· · ·

By March 1 I had surveyed ninety-four dens. I found that dens that were made shortly before dark and spontaneously left shortly after dawn contained forty to fifty-five scat. Thus, given a night of around twelve hours, the grouse produced about four scat per hour. Further, I flushed eleven grouse from their dens at various times in the day from morning until late afternoon. By plotting the number of scat left in these dens with respect to the time the birds were flushed I found that the scat numbers increased linearly: from none at 7 a.m., meaning they had very recently entered the den rather than overnighting there, to near thirty by early after-

Snow denning by ruffed grouse.

noon. In contrast, in the three dens where I flushed the bird in late afternoon or near dusk, I found only three or four scat, indicating that the grouse had left their dens and then reentered the snow either for overnighting or for brief stays before re-emerging after sundown to feed before their long overnight session in a den.

The scat clock suggested that in January and February the grouse fed mostly in the early morning and at the end of the day and spent a large part of the day under the snow. Aboveground observations were in agreement: in a month of twice-daily hour-long walks I never saw a grouse in the daytime except those I flushed out of their snow dens. Grouse sitting in bare deciduous trees feeding on buds would have been easily visible from at least sixty meters. But on only one evening (sixteen minutes after official sundown) did I see a grouse in the top of a white birch plucking buds from the tree. That is, the grouse were as if absent to any traveler through the woods during the daytime.

In addition to denning times, grouse behavior also left other records in the snow. First, most of the den entrance holes were angled grooves into the snow, so the birds dove into the snow with wings held close to their sides. The entrance depression from the dive led into a usually short tunnel of roughly the grouse's diameter, suggesting the bird pushed itself forward on its chest and belly rather than walking or fluttering. The tunnel was usually less than a meter long, although one extended to three meters. The longer tunnels often had one or two of what looked like peepholes along its length, showing where the bird had stuck its head up out of the snow on the way to its denning place, the cavity at the end of the tunnel. The cavity always had a solid base to perch on and launch from, likely formed by packed snow and perhaps also snow that had softened from body heat and then frozen.

Most of the scat was of relatively solid matter in the form of the aforementioned sausage-like pellets (1–2 centimeters long and weighing on average 1.2 grams). In a den they were piled side by side and on top of one another into a tight stack, so the bird apparently did little turning after it had settled. The pellets were not messy; even when thawed they did not soil my fingers. However, dens with large numbers of scat (near fifty, as in overnighting dens) often also contained three or four much larger fecal pellets (seven grams each) that when fresh or thawed are a sticky brown semi-liquid paste. These pellets, when present, were always separate from the large stack of scat. They did not show smear marks, as would be expected if they had been in contact with the bird, so they had probably been deposited as or shortly before the bird left the den. Total scat usually weighed about a seventh of a grouse body weight.

Most of the snow dens had no foot tracks leading in or out, indicating that the birds had entered the snow by diving into it and exited by jumping up and out at the end of their dens. As expected from such behavior, the dens were in open space, providing for free access.

Den locations were also in part related to the presence of others. I had expected to find small groups of grouse denning in close proximity, as I had commonly seen several feeding in groups at dusk in the fall. Twice I flushed two from separate dens close together, although I more often found empty dens near one another.

Walking my grouse loop and seeing the new den locations showed me that there was no single exclusive reason for the cluster denning. Paired dens had been made a day apart, and some were probably night dens that a bird had made near its earlier dens.

Given these observations, I tested whether the birds were attracted to marks in the snow that mimicked den entrances. Traveling the same loop, I made a hundred fake entrances by tossing onto the snow, and then yanking back, a surrogate for a diving grouse (a dead bantam rooster attached to a strong string). However, although the facsimiles mimicked the entrance marks of grouse dens almost exactly, no new den appeared near any of them. Apparently dens become associated with one another either by two grouse denning close together or by one grouse making a new den near its previous one. (Grouse never reuse a den.)

Originally I had wanted to know if grouse spend time in snow dens both day and night. They do. Long daily denning is uniquely possible for grouse because of their food. Unlike the other birds in these winter woods they do not need to forage continuously just to get by. Their plentifully available food, namely tree buds, can be gathered almost anywhere, probably quickly and at the birds' convenience, with almost no search. On the other hand, digesting this cheap and easy food is a lengthy process — storing it in a crop, then passing it on to the gizzard, from there to the glandular stomach, and finally to the intestines — and hence the grouse have ample time to linger in their dens. But I wondered if, in addition to serving as a place for digestion, the denning might also have permitted the grouse to evolve a strategy to reduce predation in the winter.

In the absence of snow the grouse spend most of their time hidden on the ground in dense thickets. But in the winter their food is buried under snow; they are not able to feed on the ground as they do at other times of year, so instead they find food in the tops of trees. Most of their feeding then occurs at dawn and dusk, and they have a tendency to do it in the company of others. Both of these trends — to be crepuscular and to join others — should reduce their vulnerability to predation by their main predators, great horned owls and goshawks, when they are exposed in leafless treetops where they can be seen from afar.

They also need to avoid predators the rest of the day. What better way to do that than to become invisible by hiding under a layer of snow? But there are caveats: a clever strategy almost always engenders a counter-strategy. For example, a predator might learn to track the grouse (as we can), by making deductions from signs on the snow. This in turn would be likely to produce a counter to the counter-strategy. The grouse might apply the lottery or shell-game principle: the more holes or disturbances there are on the snow that are *not* associated with a grouse in residence, the more the real signal of a grouse is degraded and becomes potentially meaningless. Of the dens I found, fewer than one in ten was occupied by a grouse, in part because grouse never use the same den twice, so that numerous decoys (from a predator's point of view) accumulate in any one area. A grouse making two dens in every twenty-four hours may leave, depending on the frequency of snowfalls, up to sixty decoys in a month. Many more depressions in the snow are created by animal tracks and by snow cushions falling off branches.

Individual grouse choosing to den near others may also forage in a crowd, so that if one is attacked the others will be alerted and able to escape. Further, even if a predator did manage to pick an occupied den, it would have to know where, in relation to the entrance, to pounce upon the grouse. This might be difficult to determine, because, at least under the condition of deep snow in which I observed the grouse, the tunnels were of various lengths and angled in different directions, so the entrance holes did not pinpoint the birds' location. Finally, a grouse's exit from a den is explosive. (My one attempt to photograph an exiting grouse that was right in front of me failed miserably: even though I knew the bird was about to appear, my finger on the camera trigger was far too slow to capture its movement.) But, as a final observation that combines all of the above potential advantages of the anti-predator advantage of the snow dens, I at no time saw signs of a predator having disturbed an active or previously used den, although coyote and weasel tracks were common in the same woods.

By March 4 I thought I was done with the grouse. Counting their scat was less "engrousing" than it had been before. I needed to walk the route once more and time myself in order to be able to estimate the length (five kilometers). It was a sunny noontime, and temperatures had soared to 35°F. Hairy and downy woodpeckers drummed, and blue jays held their always noisy first spring convention. An even surer sign that winter was winding down was the snow fleas. It was the first day I had seen them on the snow. Two moose, a cow and her calf, sauntered ahead of me, and I saw them for a brief few moments. But none of these was as exciting as the sight and sign of three grouse doing something they had not done for at least the preceding two months.

A couple centimeters of snow had fallen in the night, and I found one fresh grouse track that led into a spruce thicket, where the bird had hunkered down without covering itself with snow and had left sixteen scat. It had come there that morning and stayed awhile, but had left long before the early-evening feeding time. I flushed two more grouse, which also had sought shelter for a long part of the day but had not buried themselves even though the snow was soft and fluffy. Instead they had made molds in the snow under low-hanging branches of balsam fir. Soon after this, they no longer overnighted in the open areas of the deciduous woods and again roosted, as in the rest of the year, several meters high in the thick branches of spruce and balsam fir trees growing in thickets under which their scat was scattered over the ground.

The overnight change of grouse denning behavior, a reaction to changes in both snow conditions and temperature, was a gift to find so unexpectedly. It was another reminder of the rewards of continuity of observations — of one bird behavior at a time.

13

Crested Flycatcher's Nest Helpers

A FOOTLONG PIECE OF HOLLOW LOG WITH A HOLE IN ITS side, and with two pieces of board nailed onto the top and bottom, is a great bird house. One that I made of a piece of black cherry from the winter firewood was used by a pair of great crested flycatchers. The previous year they (or another pair) had occupied a bird box of boards that was meant for wood ducks, until a pair of grackles used it in early April, when the beaver bog in Vermont had drained and they could no longer build nests directly over the water in cattails.

Now, on April 25, 2010, the leaves were popping out, and serviceberry, trilliums, trout lilies, and lilacs were in full bloom. And, oh, the birds! Migrants were still coming through, mostly white-throated sparrows and ruby-crowned kinglets. I heard the first warbler, saw a chickadee slip into the bird box that tree swallows had nested in the year before. The broad-winged hawk was back, and the snipe displayed over the bog. The phoebes had already built their bulky nest of mud and moss, this year attaching it to a narrow "ledge" of board under the roof and over a window. The neighbor's bluebirds had just laid their first clutch of five eggs, and I heard a woodpecker tapping and discovered a sapsucker making phloem taps in a pine tree, the only time I ever saw them do that on a pine.

May 4, 2010. I found the first blue eggshell of a robin on the ground. Just-hatched Canada goose goslings, as yellow as the carpet of butter-colored dandelions now on lawns, were visible on a neighbor's pond. Purple and white apple blossoms adorned the trees. I checked the bird box where I had seen the chickadee slip in the week before. Lifting off the cover, I saw nothing but fluff almost to the top, but teasing it apart revealed two speckled eggs. So the female had just started to lay her clutch; she would hide the eggs daily after laying her morning egg and before leaving for the day. The northern orioles had just returned and were already starting to build their hanging nest from fibers pulled out of the stems of last year's milkweed, and a pair of kingbirds were attempting to steal the fibers to use in their own nest. Tree swallows stopped by. But there was yet no sign of the great crested flycatchers that had nested here the year before.

May 8, 2010. At noon it was snowing and chilly (26–28°F) under a dark sky. Every two steps forward toward summer were still being followed by one step back to winter. No tree swallows were in sight: they could not find food on the wing on a day like this.

May 11, 2010. It was still a chilly 25°F at dawn, but now under a clear blue sky and the rising sun the air warmed up quickly. Some of the frost-sensitive leaves had been killed. Those of birch, maple, and cherry would revive. The swallows were back, and the orioles' nest looked finished. I was glad to see the phoebe fly to its nest. When I tapped on the chickadee box and lifted the cover the incubating female flew out, and within the insulating and concealing fluff I saw an exquisite, tiny, deep round cup cradling six eggs. The clutch was now full, and the eggs' volume and weight exceeded those of the female that had laid them.

May 17, 2010. Finally! I heard the ringing calls of great crested flycatchers and noticed a pair examining the nest box I had made from the hollow black cherry log. As one bird entered the box the other flew to the nearby apple tree and called loudly.

From then until June 19, despite being near the flycatchers' nest box every day, I scarcely mentioned them in fifty-three pages of dense notes about five nests in view — tree swallows, phoebes, orioles, chickadees, and the crested flycatchers — except to note that when I tickled their nest box on May 27 the incubating bird snapped its bill in warning or alarm and then flew off. On June 19 I merely mentioned that the flycatchers were perching conspicuously by the nest entrance and that they seemed unperturbed when I opened the nest box and saw the now partially feathered young hunkered down. They too snapped their bills in some kind of warning display; the sides of their mouths were white, not yellow as in most birds. But then, starting the next day, I took twenty pages of notes on the crested flycatchers in just three days.

The flycatcher story started the morning of June 20. At 7:30 a.m., on my usual morning bird walk with what was then my regular entourage, two dogs, two Canada geese goslings, and a rooster, I noticed the male and female flying back and forth from one treetop to another. They kept their distance from each other and flew slowly, with deliberate shallow fluttering wing beats, and with their tails slightly elevated as though acting as air brakes. The performance looked like a display flight, and at the least there was a hubbub, but I detected nothing unusual at the nest. Crested flycatchers are normally noisy near the nest, and were being noisy now, but seemed to be making themselves even more conspicuous than usual. Why? What was going on?

As I made myself comfortable in the grass to settle down for a long watch, the dogs and the goslings lay down beside me while the bantam rooster scratched in a raspberry patch. I pulled a folded sheet of paper out of my back pocket and started scribbling notes that I would later transcribe over a cup of coffee at the kitchen table.

If the flycatchers' commotion was odd, I soon noticed something even more surprising: one of them often followed another. This late in the breeding cycle the pair would usually stay separate, independently foraging in the woods to feed the young, rather than carrying on with each other near the nest. Could there be, as I thought I briefly saw, a *third* bird? Soon there was no doubt about it — three flycatchers were flying together. Had the young morphed and fledged overnight? That prospect was remote, but I had to eliminate the least likely possibility first, so I got my stepladder and, as I had done several times before, climbed up to lift the lid on the nest box and peer in. As expected, the five young were still there. Again they hunkered down, and again one snapped its beak at me. The three adults, which had been paying very close attention to one another for at least a half-hour, ignored me.

An hour later the adults finally started carrying food to the nest box. Curiously, though, even when one arrived with an insect in its bill, it would linger for minutes at a time, perching on a vine by the box while scanning in all directions, before slipping into the nest box to feed the young.

Note taking, after three almost full days of continuous watching, was becoming increasingly tedious, but I needed to record as much information as I could, because I knew nothing about what was going on and therefore could not tell what might be important. The process of getting data, mining inferences, and comparing hypotheses to try to solve a puzzle is seldom predictable and almost never straightforward. With luck a pattern emerges so one of several hypotheses can click into place. Finally I realized that some of the data I had collected was relevant to judging a hypothesis. My first inference was that the excitement was due to the sudden presence of a third party, another great crested flycatcher. But why would a third bird have arrived, and why would it stay? Did it want to usurp the nest cavity for its own use? That didn't make sense: What would it want with a nesting place near the end of the breeding season? Also, if the bird was an intruder, the pair would surely attack it, and I had seen no attacks. I doubted my own eyes: Was I really seeing a third bird interested in this nest? Maybe it had simply been attracted to the commotion, or to something else, such as the red-winged blackbirds that had recently converged in a meadow. The redwings had focused their attention on the ground, and as I ran over to investigate, a mink ran by my feet in the thick grass. So the riddle of the blackbird crowd was solved in seconds. I had also just seen a mob of birds—robins, cedar waxwings, kingbirds, and catbirds—flying together. They were chasing a blue jay caught in the act of raiding a robin's nest in the birch tree directly in front of me. In these examples the causes of the disturbances were obvious. But for the flycatchers' unusual behavior I could see no cause, and nothing made sense. By now I was convinced, however, that there were three birds, or possibly more since I could not distinguish them as individuals.

The parents, which came and went hunting insects for the young, brought primarily dragonflies, of at least three species. On that afternoon they delivered five dragonflies in three hours. I had the impression that they made fewer trips to the nest than before. But impressions don't count. Numbers do.

I was out to watch the flycatchers by 5 a.m. the next morning, and saw the pair. But a half-hour later a second *pair* flew in, and almost immediately there were chases, during which I heard the high-pitched staccato calls that were different from others I had heard before. These were all-out chases, but the chased birds were unwilling to leave. One pursuit culminated in two birds' tangling in midair and then falling straight down in a fluttering ball and landing in a dense patch of goldenrod. No more were the flycatchers merely shouting from the treetops. Instead, they were duking it out after what had apparently been a long test or buildup period. But what was the quarrel about? Were the newcomers a second pair wanting to kill the young and take over the nest box?

At 8:03 a.m. I got another surprise. A flycatcher arrived with a dragonfly, perched at the nest box, and called. A second bird answered from the top of a neighboring tree. The first bird entered the box, emerged without the dragonfly, and called again, then the two left together. All was quiet for a minute or so. But then suddenly a single bird, which remained silent, flew in with something in its bill. It too landed next to the box and hesitated, and while I was trying to identify its prey (a beetle?), it flew off, still silent, without having entered the box. Just as it flew into low bushes, two other adults appeared as if out of nowhere. Neither of these two carried food, but they called loudly. It seemed as though the previous bird had tried to escape them. Was the pair trying to repel a bird that had come to help them rear their young — to be a helper at the nest? The idea seemed preposterous. However, my subsequent observations eventually left no doubt that one or both of the newly arrived birds were not only trying to feed but actually succeeding in feeding the others' young.

The resident pair generally stayed together as a couple, with one perching high in the trees and calling often while its mate flew low into the woods and bog to hunt. Whenever one returned with food, the other perched high and called while the first landed on the grapevine in front of the nest box and called, then entered the box with the food. The pair always maintained a conversation" by calling back and forth. One entered the box, fed the young and came out, stayed awhile and called, and then the two left together. Often soon after their departure a lone flycatcher showed up with an insect in its bill, slipped quickly into the nest box, emerged minus the food, and left. It did all this in silence, although another flycatcher called from a distance.

The second two flycatchers (rather than just one, as I had by now thought likely) came mostly when there was silence around the nest. Occasionally the two pairs' visits overlapped and violent chases ensued. But the chases gradually became milder; the pairs seemed to reach an accommodation, perhaps mainly by avoiding each other. The net prey brought in per hour increased from just one insect on the first morning to eight, twelve, seven, fourteen, and then eighteen on subsequent days.

By the third day, when I could by behavior distinguish the residents from the others, there were no more fights and no more chases. The second female had become a true helper, whereas her mate hung back from the nest. Of one day's total of eighteen insects, she brought in ten. Perhaps the parent birds tolerated her by then because she was useful, or perhaps (as I think is more likely) they had simply become habituated to their helpers.

As usually happens, answering one question leads to another. The obvious next question was why a second pair had arrived to help raise another pair's young. Given what we know about birds we can make educated guesses, and my guess was that they were neighbors whose nest had been destroyed and who were playing out their parental instincts on the young of the other pair. The parents at first perceived their interest in the nest as an intrusion but eventually became used to them, and/or the second pair became more skilled at approaching the nest. But additional and almost inadvertent observations made after all the young fledged on June 23 suggested another possible twist.

With the young gone, I could examine the nest. It was clean, so the scat must have been carried out as soon as it was produced. On a whim I pulled out the now-useless nest to see what it was made of, and was surprised that it consisted almost solely of the long dry needles shed by white pine trees. Curiously, there was an intact great crested flycatcher egg under the top layer of the nest. This was strange, because birds generally do not cover up their own eggs unless they leave the nest for a long period. When I put the egg in a bowl of water it floated; it was therefore either spoiled or incubated. Once I opened it, some still-intact yolk and a lack of blood vessels made clear that it had not been incubated, but was spoiled.

The rejected egg interested me greatly as a possible clue to the riddle of the shared nest duties: it could have belonged to a pair other than the one that fledged young. Perhaps the helping pair had been at the nest box first, long enough to produce one egg, after which the now-resident pair had arrived and driven them off. The new occupants would have gone through their whole nest-building repertoire regardless of what was in the nest at the time; they would have covered up any existing egg or eggs as though they were debris. The evicted pair would then have been without a nest hole to start a new family (suitable nest holes are often rare). Perhaps they did not find one and eventually came back to check their former nest, only to find ready-made babies.

The day after the young fledged, the adults no longer called near the nest box, but at times I still heard both their demonstrative calls and the young's begging in the distance. However, five days later, at 9:35 a.m. on June 28, two adults arrived. One chased the other for at least a half-hour, making staccato rattling calls. The chased bird did not want to leave.

The flycatcher parents' attempts to keep the interlopers from caring for their young was most likely related to the avian evolutionary strategy of reducing nest parasitism, while the other two were probably motivated by a strong drive to feed their own young. Both pairs were following their usual adaptive scripts, which in this case were off the mark because of changed circumstances. A friend told me about a similar situation: a pair of house finches (*Carpodacus mexicanus*) nesting in a cranny on her porch had a nestful of young, and a pair of English or house sparrows (*Passer domesticus*) were not only feeding them but also carrying off their poop, whereas house finches normally let the wastes collect where they may. The reason this switch in behavior may at first seem dubious is that house sparrows are well known to viciously fight other birds at nesting sites, evict the others' eggs and young, then lay their own eggs and raise their own young there. Why had they not done so here but instead become apparent altruists?

The likely answer is the same for the sparrows as for the flycatchers: they were expressing their parental instincts after their own nest and young had been destroyed. This parenting instinct is so strong that it can be uncritical and can be misdirected, as when it induces a small warbler to feed a parasitic cuckoo nestling that may be several times its own size and looks nothing like its own young.

14

---≈---

Red-winged Blackbirds
Returning

THE CLASSIC DEPICTION OF THE RED-WINGED BLACKBIRD IS one of male fighting, dominance, territoriality, polygamy, harems, and the females' attraction to resources or to the males that have them. But this bird also has popular appeal through its showy conspicuousness and almost universal familiarity. In northeastern America redwings are one of the first bird species to return in the spring, and many springs I have eagerly awaited their arrival at a beaver bog in Vermont. The males usually appear quite suddenly on sunny days when the snow starts melting in the first week of March.

In early March of 2011, I was expecting to see male redwings show up at the beaver bog next to our house, but instead, at dawn on March 7, two males were perched in a spruce outside my study window in a white-out snowstorm, and within minutes they were picking up sunflower seed spilled from a birdfeeder that during the winter had been visited only by chickadees, nuthatches, and blue jays.

Every spring, red-winged blackbird males arrive in a group, usually appearing a month or more before any females. In previous springs the males had also fed on sunflower seed, and had flown back and forth between the bog and the feeder (females never came to the feeder). But these two did not fly into the bog. They continued to shuttle between a red spruce tree by my window and the adjacent snow-covered ground under the feeder. Perched under a branch covered with a thick cushion of snow, using it as a roof, they made an attractive picture that induced me to watch and sketch them. They were excellent models, giving me their whole day and wearing their finest garb, their breeding plumage. But unlike the popularly depicted male redwing plumage, theirs showed not the tiniest hint of bright flaming red. They hid their red badges of maleness under their black wing feathers, leaving visible only the light yellowish edge that bordered the red.

The two remained silent all day in the continually falling snow, shuttling between the feeder and their shelter under the spruce bough, where they perched fluffed out and side by side. Occasionally one flicked his tail and stretched a wing. When one flew down, the other immediately joined him, and when he came back up to the perch he again sat next to the other. They stayed all day, until an hour before dusk they finally left together.

Nighttime temperatures dropped to 0°F that week, and it was not until the evening of March 13 that I finally saw a flock of about thirty male redwings perched in a tree several kilometers from our house. The next morning males had arrived in our wetland and were yodeling again and again, producing their cheerful-sounding *oog-la-yee* territorial advertisement. The snow had then been melting for several days but was still a meter deep in the woods. Crows were cawing, and groups of them came flying overhead in a northerly direction.

By the end of March, when the male redwings in our wetland in

Vermont had long since established their territories, others were still on the move elsewhere. On March 26, as I was heading east to Maine, I met a flock of twenty males at a spot near Mt. Washington where there is almost unbroken forest in all directions. These redwings were passing over the road in front of me, flying into a strong wind from the northwest. Their resolve to buck this wind and fly as a group to an apparently agreed-on place, which at this time would be a wetland to nest in, hinted at unanimity of purpose, not aggressive rivalry. How had these birds met and become a flock? Given the two males I had seen previously that acted like close friends, and given what I had observed in my casual watching of redwings in earlier years, it seemed not impossible that these twenty or so individuals could be a *community* of birds, perhaps returning to a common home area they had shared before and would share again. If so, how would that jibe with the classic story of male dominance and territoriality?

Red-winged blackbirds are close relatives of grackles and orioles. They belong to the family Icteridae and are one of the most social of songbirds. Social behavior is part of their genetic heritage. In our beaver bog wetlands, the common grackles come as a flock, stay as a loose flock, often nest near one another, and in the fall aggregate into flocks of thousands that sweep through the forest like huge rolling waves. To varying degrees, the different species of icterids form communal roosts of thousands of individuals in the non-breeding season, and some, like the Oropendolas (a group of several genera of South and Central American icterids), nest in colonies.

Long after making my own observations I read two monographs on blackbirds. The first, by Gordon H. Orians of the University of Washington, dated from 1980, when the scientific literature on the topic amounted to about 170 works. The second, a 1995 report by William A. Searcy of the University of Miami and Ken Yasukawa of Beloit College that focused almost exclusively on red-winged

A red-winged blackbird male displaying its red epaulets.

blackbirds (*Agelaius phoeniceus*), listed 440 scientific publications. The authors stated: "The behavior of red-winged blackbirds in the wild has been studied as extensively as that of any species of bird in the world."

The reason for the abundance of studies is simple: redwings are one of our most common birds, and they are highly visible because they live in open habitat, such as cattail marshes and wet fields. Redwings populate almost the entire North American continent from Alaska through Central America and from the Atlantic to the Pacific. And because of their wide distribution and conspicuousness, almost anyone can have access to them.

These older publications, with anecdotes and observations not focused on current theoretical problems, made what I saw interesting, at least to me.

During the breeding season the redwing males are, according to these monographs, "strongly territorial" (Orians) or "classically territorial" (Searcy and Yasukawa); they reside preferentially in cattail marshes, where they occupy plots "more or less exclusive" of other males. Redwings are one of the few songbird bird species that are polygynous (several females may share the same male or, as conventionally stated, "males have more than one female"), whereas the majority of birds are socially monogamous. A female chooses a mate indirectly by choosing the territory that a male holds. These females also mate with males of other territories, but not with floater males, which do not have territories. Males do not build nests or incubate eggs, and for the most part don't feed the young. But they are fierce defenders of the home turf. As Searcy and Yasukawa put it, those who study redwing breeding behavior are "familiar with the sensation of being struck on the back of the head by a male who has just performed a power dive." I've never experienced this myself, but then I have also not studied territoriality, though I've spent much time in their territories.

Redwings' territorial behavior has been experimentally examined using male redwing taxidermy mounts set up in redwing wetland. The results are dramatic: territory holders attack these mounts, rip them apart, even tear out their eyes. Despite the seeming consistency of the classic picture, what I had seen in my wetland in Vermont left me scratching my head. Context is obviously important, if not everything, but it still seemed incongruous to me that males who traveled back to their home bogs in small groups, apparently tolerating one another, would become enemies after they got home. In the days after the males returned to the bog they spread out over the bog, but in the evenings flew together to a roost. In their daytime displays from the tops of bushes or cattail stalks, they were at times within several meters of one another, vocalizing and prominently displaying their red wing badges, but without any apparent interactions despite close proximity to each other.

The females come back a month or more later than the males. They are cryptic (camouflaged) and easy for the human eye to miss, but the males are alert to them. I often saw one female with four to six males in swift flight behind her. The males' intense pursuit and the female's equally intense evasive maneuvering looked like an act of aggression in progress, but these chases did not end in aggression. According to the classic picture the female could have been shopping around, examining the then-established male territories to find the one with the best resources, before deciding where and with whom to settle and raise a family; and when she entered a male's territory, its owner should have courted her, and perhaps even offered food, to impress her with the resources to be found there and to lure her to stay. What I saw did not match this picture. These females acted as though they were doing their best to get away and hide in the dense low vegetation.

Could the males' pursuit of the newly arrived females be related to mating? Perhaps, but I had never seen either copulation or fights at the end of such chases. Supposedly the females were looking for territories to settle in, and after they chose, the territory holders would mate with them. But the timing was off: birds normally mate at or just before egg laying, which would not begin until three to five weeks later. And when I did see copulations later, they were not preceded by chases. Thus, neither the males' territorial behavior nor the females' responses to the males seemed to fit the expected schedule.

That said, I also saw spectacular male-on-male chases. But rather than in March when the males arrived at their territories, these occurred in late May, when eggs had been or were being laid. Some went round and round the entire bog. According to theory, the "fittest" males secure the best territories, and in deciding where to settle a female chooses not only the direct benefits of occupying the best territory but also, indirectly, the male's fitness. So it pays a male to fight other males and exclude them from his territory. But these chases extended across the territories of several males.

The scheme of territorial defense makes intuitive sense and has excellent experimental support. For example, as previously mentioned, when researchers inserted a stranger in the form of a stuffed male mounted on a pole into a territory, the territory holder attacked the interloper. This is convincing evidence of territorial defense, but using a stuffed male for the test automatically introduces an *unfamiliar* potential rival, one that the resident males will not recognize as a neighbor. Resident males in the marsh often stay several meters apart and pay no apparent attention to one another — and these are territorial males that are displaying their red epaulets! The fact that even at the end of May, when the females were completing the laying of their clutches, I sometimes saw a group of males violently chase another male hundreds of meters into distant territories at another end of the marsh suggests male cooperation among at least some neighbors.

From the studies summarized by Searcy and Yasukawa we know that redwing territory holders return to the same territory in successive years. This may explain the lack of aggression I saw among just-returned males: the individuals may know each other and may have already established boundaries. They may be neighbors that have traveled together when outside the home marsh. The males they chase out of their marsh may be the floaters, who are complete strangers. However, in the chases after females that I observed, the object of the chase never left the marsh but instead appeared to try to hide in low vegetation, and the reason for the chasing was more obscure.

· · ·

Ironic as it may seem, as intensity of competition increases, one of the standard evolutionary solutions to it is more cooperation. We humans are a prime example. Local squabbles exist in any group and in any relationship, but a common threat unites by the common interest it creates, and can cause differences to seem less important. A main factor making cooperation possible is that the individuals know one another or at least know they share some identifying characteristic (such as the specific scent of a particular social-insect colony). Red-winged blackbirds can distinguish among individual humans — as researchers found out when the birds preferentially attacked those people who "raided" their nests (to band the young).

Many of the nests in the beaver bog I later found prematurely empty. I usually did not see the raiders, but the redwings would have, and having identified them, they regularly flew up out of the marsh to attack ravens, crows, and any passing blue jay, bird species that routinely take the eggs and young of other birds. Specific humans who have threatened their young receive the same treatment.

Cooperation among the male redwings was often evident. On the evening of June 8, 2007, I was surprised to see two otters playing in our beaver pond. The next morning they were in the bog next to the pond, and a mob of red-winged blackbirds and grackles (both of which had eggs and young in their nests in the cattails) were diving at them and making a huge ruckus. The mixed-species mob advanced across the bog, keeping up with the otters below. The male redwings from most if not all of the marsh had aggregated within perhaps a dozen square meters. All the birds were scolding loudly and taking turns diving at the otters, thereby notifying others that the intruders represented a hazard. On another occasion I witnessed the birds mobbing a mink. I sometimes lost sight of their target as it was obscured by vegetation, but I could easily plot its progress through the marsh by watching them keeping pace with it. At no time during these episodes did one redwing male attack another. The No Trespassing convention had been waived in the face of communal danger. The red-winged black-birds' behavior illustrates what behavioral ecologists have come to call the "dear enemy" phenomenon, in which territorial residents display less aggression toward familiar neighbors than toward strangers.

15

Phoebe Seasons

WHEN JOHN JAMES AUDUBON WAS COURTING HIS FUTURE wife, Lucy Bakewell, near Mill Grove, Pennsylvania, the two became fascinated by a pair of eastern phoebes nesting in a cave. They often watched these relatively tame birds, and Audubon wrote that the sight of the freshly laid eggs was to him "more pleasant than if I had met with a diamond of the same size," and that it "filled my mind with as much wonder as when, looking towards the heavens, I searched, alas! in vain, for the import of all that I saw." The eggs hatched, the young grew up, and Audubon (presumably with Lucy holding them in her hands) tied silver thread onto their legs. The next spring (1803) he found two of the banded birds returned from their winter residence. He is credited with the first recorded bird banding in America.

Audubon's words seem personal to me. Just as his early love of the natural world was colored by phoebes (then known as pee-wee flycatchers), my history was heavily influenced by another flycatcher, the pied flycatcher (*Ficedula hypoleuca*), a pair of which nested in a bird box on my family's cabin in the Hahnheide woods in northern Germany, where I lived six years as a young boy. The birds caught my eye, and I experienced a soothing feeling of comfort to have them so close, almost living with us. And when eventually the light blue eggs appeared they evoked the same kinds of emotions Audubon described. Then when my family came to Maine, finding the phoebes nesting here immediately made this place our home. By that time they were much more common than in Audubon's day because of an increase in the number of man-made nesting sites. They were nesting in our barn, our outhouse, and our shed. Much later, phoebes nested on our house in Vermont, and they were and still are a fixture at my new home in Maine. Phoebes invoke the joys of spring and domestic harmony, and although they are not scientifically classified as "songbirds," their vocalizations bring me more cheer than those of any canary.

Arriving back at our woods in Maine on April 21, 2012, we expected to see the (not necessarily always the same) pair of eastern phoebes near their nest on a log above a window and under the roof of my original cabin, where I had lived seasonally before we moved into the new cabin fifty meters away. The ponds were still frozen over, and the steep path up to the cabin was covered with thick ice with meltwater gushing over it. As usual at this time, the snow in the shaded woods was still crusted in the morning and hard enough to walk on. The yellow-rumped warblers were migrating through and visiting the first flowering red maples, possibly for nectar or small insects. Blue-headed vireos were singing and tree swallows were fighting for a nest box. In the evening a woodcock displayed on and over the clearing, and later at night a grouse drummed in the moonlight.

To my great joy and comfort, a pair of phoebes that I considered old friends were back. The next morning one of them perched beside my window, swiveling its head and flicking its tail up and

A phoebe and nest containing four phoebe eggs and one cowbird egg.

down in the phoebe's singular salute. Occasionally it sallied to the ground to catch an insect, then flew up to perch in a birch or a sugar maple whose branches brushed up against the cabin. The birds gave an occasional soft *cheep*, but not once did I hear them make the typically loud and emphatic *phee-bee* and *cheer-vreet* calls, which often alternate in the males' territorial advertisement songs. I expected them to start nesting shortly. However, for the next two weeks the weather was cold and wet. No insects flew, and the phoebes and many other birds left or became silent.

Nesting activity resumed or began on May 7. The tree swallows were carrying dry grass into their nest box, and the phoebe pair examined the old nest site on the old log cabin where a pair of phoebes had nested every spring since it had been built twenty-seven years before. But the phoebes also flew in and out of the cellar space under the new digs. They had not yet started nest building but were enthusiastically inspecting possible locations.

After another of their forays under the new cabin, one of the pair flew high into the air, rising above the tallest maple trees at the edge of the clearing, performing what looked like a poor imitation of a woodcock sky dance. Immediately afterward they began carrying mud onto a small board shelf I had tacked to a beam in the space, the "cellar," under the floorboards. The female made trip after trip, bringing first mud and then moss from the edge of my rock-lined well about a hundred meters downslope. Her mate accompanied her on these trips but did not carry anything himself. The nest was finished by May 19, and two days later it held two eggs. But to my surprise I saw a phoebe inspecting the area under the eaves at the old log cabin as though still checking for potential nest sites. Phoebes are territorial: two pairs could not live this closely together. Subsequent observations deepened the mystery.

I had been able to watch the pair closely during nest building and egg laying because the opening to the cellar was conveniently below my window and because they perched nearby, on the still-bare branches of a white birch tree three meters from me, or on the tops of dry mullein stalks in the clearing just beyond the tree. At this time, well into nesting, they were usually not vocally demonstrative except at dawn.

The next day, when she added the third egg to the still-incomplete clutch (the typical clutch is five), I heard extraordinary, animated vocalizations. The male appeared to be "singing" as phoebes usually do at the beginning of the nesting season; he made two kinds of *phee-bee* calls but, oddly, again performed a sky dance over the cabin and the trees while making them. Unable to think of a reason for this sudden display in the normally rather secretive egg-laying period, I checked the nest that afternoon. I reached in and felt with my fingers — the nest was empty except for some eggshells!

Throughout that day and well into the evening the male called repeatedly and continued his fluttering display flights over the clearing. I had heard and seen such activity in early spring, and once on the day the young fledged, but never at this stage. Since the nest disturbance I had seen only one bird, never two at once, and I realized that the female was missing — probably killed on the nest by a predator that had also destroyed the eggs — and that the male's continuous calling and his sky dancing had to do with her absence. Flycatchers are classified as suboscines, a moniker that distinguishes them from the oscines or "true" songbirds. They are not melodious like thrushes and warblers, but I'm not sure why their vocalizations, given in the same contexts as those in which oscines sing, are not considered songs, regardless of their subjective tonal effect on us. What emotion may prompt their utterances is not for us to say. I suspect that, for them as for us, context matters: we may cry when we are hurt or sad but also shed tears of joy; we may scream in terror or in rejoicing.

The next morning we awoke before daylight to the phoebe's loud and constant calling. But as the calls continued all morning and well into the afternoon, they became muted. Finally at 4 p.m. the calls stopped.

On the third day the male at first stayed around the cabin and still called from the treetops, then frequented low perches and started to capture one insect after another. I wondered: Might he be fueling up for a long trip? At the next dawn the silence felt deafening, and we did not hear or see him again that summer.

Fall and winter passed. Hoping the male might return, we awaited his arrival anxiously in early spring of 2013. Would he come with a new mate, or would he come alone and advertise for one here?

The year had so far been unusual: we had had little snow in the winter, and instead of a long cold wet period in the spring there was a heat wave. In mid-March temperatures reached near 30°C and all the snow melted. Phoebes are one of the earliest migrants to return, and I thought they would come even earlier this year.

On March 19 I saw a lone phoebe. It was silent except for occasional *cheep*s. I expected a potential mate to arrive soon and nest building to begin.

Two mornings later there was still only one, almost silent, phoebe around the cabin. At 1:10 p.m., though, he suddenly erupted into nonstop song, making the *phee-bee* and *cheer-vreet* calls alternately, with an occasional lower trill. After eight minutes of calling he caught a large insect, pounded it on a branch, and swallowed it. After that, without making any more calls, he left, flying high into the air toward the east. I expected him to come back soon.

The weather then made an abrupt turn. It snowed, and temperatures on March 26 were below freezing in the daytime and far below freezing at night. Winter was not over. However, he or another phoebe had returned just before it snowed. He was not vocalizing and stayed near the ground. There could have been no insects to catch on the wing.

By March 30, after several centimeters of fresh snow, the bird was gone, and by April 7 there was still no sign of a phoebe at the cabin. But nine days later I saw two of them. I could not tell if they were male or female from their appearance, because to our eyes the sexes look identical. One repeatedly checked the old nest site under the cabin roof. It fluttered near the tattered old nest, landed there, and made soft churring sounds. It landed on another ledge under the roof, then flew back and forth between the two, each time sweetly chittering as if excited. The second bird hung back and kept some distance from the cabin, but whenever it came near, the first immediately fluttered to the potential nest sites as if to either draw attention to them or defend them. Sometimes one flew after the other.

The next dawn a lone phoebe continuously broadcast his song from the treetops. Around 7:30 a.m., when another phoebe arrived, he stopped singing and again visited the nest sites. He would seemingly squat on each site for a few moments, then fly to a nearby maple twig, shake his wings and wag his tail vigorously, and flit back to the nesting spots. Sometimes he suddenly flew from the nest area and attacked the other bird. Once the two tangled in flight and dropped to the ground before separating. Several times one chased the other in fast flight through the woods, but quickly came back to repeat the behaviors at the potential spots for placing a nest. Their standoff continued all morning and into the early afternoon, with one squatting at the nest sites and attacking whenever the other came near. They were rivals fighting for possession of a place to nest in this territory. By 2 p.m., one male was singing vigorously and still flying to and from the possible sites, but the other had left.

The next morning, April 18, with the other bird apparently chased off, the victorious phoebe devoted less time to site visiting and much more time to attention-getting singing from conspicuous high perches in the tops of the trees near the cabin.

At 7:25 a.m. another phoebe arrived. This one was unobtrusive but remained near the cabin. The then-resident male instantly left his singing perch and hovered at one potential nest site after another, choosing spots under the roof overhang where there was sufficient ledge to attach a nest. Next he landed on the cabin roof over one site and then the other. He made excited-sounding short *chip* calls and *cheep*s and buzzing calls, and sometimes sang from directly on a nest site, something I had never seen him do before. He carried on crazily around the roofed sites. Never once did he fly *at* the visitor, nor did either bird attempt to attack. I gathered that this newcomer was a female, and that he was showing her where she might choose to build a nest and thereby become his mate.

She seemed unimpressed, though, and after less than a minute flew away toward the east. There had not been one aggressive interaction. Shortly after she left he too flew in that direction, and soon I heard him singing in the distance, far from the nest sites. He came back in a few minutes, aggressively swooped at a yellow-rumped warbler, and resumed singing around the cabin.

Twenty minutes later I saw a phoebe swoop into the cellar space under the new cabin, where the phoebe nest had failed the previous spring. In the next three days he sang almost all day, and on the evening of April 21 he again flew far over the pines to the east, where the female visitor had flown. The next morning I heard him in the woods in that direction. He came briefly back to the cabin, sang several times, and aggressively chased a pine siskin, a tree swallow, and a black-capped chickadee. His chasing of these birds that he could not have mistaken for phoebes and that could not have been his rivals might have been caused by deep frustration.

On May 10, 2013, after an absence of more than two weeks, I arrived back at the cabin in the afternoon, eager to hear the phoebe. Would he finally have a mate? It was soon clear that he did not. There was no new nest in any of the spots where one could have been. A phoebe still sang all day long as before. In the evenings he flew to the tiptops of the tallest trees around the cabin and from there made sallies over the forest, calling and performing his aerial display. In the mornings he flew under the cabin roof to the same potential nest sites, as though showing them now to an imaginary mate. He even visited the site of the failed nest in the cabin cellar.

He seemed as enthusiastic as ever, but neither mate nor rivals arrived. And so he persisted every day through May. By then a local pair already had half-grown young and phoebes in Vermont were into their second nesting. Still, every morning the lone male started to sing before daylight, at around 4:30 a.m., and continued without a break for about an hour. Then he always made his rounds of the nest sites as though showing them off, but no other phoebe was ever there to see them.

In the first week of June he was checking out new sites, as if the ones he had been examining might have been inadequate. For a test I put up a small board under a shed roof that could have been an excellent nest site. He quickly found it, perched on it, and fluttered and chittered again as though talking to a phantom female. But everything remained as before. On June 16, a warm day when I had left the cabin door open, he flew in and seemed to be searching inside; phoebes often build their nests in sheds and barns. But he didn't stay.

Finally, at the beginning of June, he appeared to give up; there was no more singing. But then on June 24 he inexplicably let loose with excited song in late evening, at 8:45 p.m., when it was just getting dark. He had not sung once for a week, and now he sang without a break for sixteen minutes, alternating the *phee-bee* and *cheer-vreet* calls as he had when he first arrived in the spring, and did not stop until it was too dark for me to read my watch dial without a light. It was his last serenade.

The next morning I heard several single calls, then no more calls at all. He stayed around for a while longer but, being silent, was not always easy to find or recognize. His tail-wags were now faint and infrequent.

On July 3, I saw him seemingly slumped down in the sun, perched on a pine stump in front of my window. Sometimes I also saw him perched briefly on the dead branches of the black locust tree in the clearing, all the way into September. Being still here, I expected him to come back the next spring to give it another try, after spending the winter in perhaps Georgia.

The first phoebe the following spring, 2014, arrived at the cabin on April 7. It was silent, so I could not keep track of it, but in any case it did not stay nearby. On two more occasions I saw it or others that made no fuss except an occasional soft *cheep*. They were near the cabin but paid no detailed attention to it or to potential nest sites. I suspected they were vagrants traveling through to other destinations. But finally, on April 22, came a phoebe that showed entirely different behavior.

I heard him in mid-morning, after a night of warm rain and temperatures around 8°C. He was singing loudly, giving the typical *phee-bee* and *cheer-vreet* calls at the rapid pace of one per second. He perched at the tiptop of a sugar maple, near the ledge under the cabin roof where the nest had formerly been (it had fallen off over the winter). He repeatedly flew down and spent minutes fluttering and perching on that precise spot, then resumed singing from the tops of the tallest trees at the cabin. He had no apparent preference among sugar and red maple, pine, oak, and chestnut; it was *height* that he sought. He most often faced away from the clearing when he sang. At 9:45 a.m. he flew west over the forest and sang at about a hundred meters' distance. He returned at 10:04, sang briefly, and then left again. Having chosen a nest site, he was now trying to lure a mate to it. But would one come?

I saw only one phoebe off and on near the cabin for the next several days, but by May 1 a *pair* was present. By May 7 the two had started a nest. They mated on the 14th and finished a clutch of four eggs on the 26th, and in the afternoon of June 28 all the young fledged.

The next year, 2015, the first phoebes arrived on April 16, and a pair was present from the beginning. Both male and female inspected nest sites from that first day, appearing to be undecided as they flew from one to another, always the same ones day after day. But finally on May 1 they started building, at the same site where the nest had been in previous years. The fourth egg of their clutch was laid on May 13.

The behavior of the two pairs differed sharply from that of the lone bird without a mate: with only the one bird present, there was singing from the tops of the trees throughout the spring and summer. With the paired birds I heard none of this singing from the treetops and few loud vocalizations at all after nesting began.

I had not expected to be able to observe how a phoebe would behave when it lost a mate, or how it would act to secure a new one. Seeing this process from up close and watching it unfold did not bring me much understanding but did bring empathy. The eventual success of the nesting pairs retrieved the joy I had felt living with similar flycatchers in my childhood.

16

———— ≈ ————

Evening Grosbeaks

THE EVENING GROSBEAK IS ODDLY NAMED. IT IS NO MORE associated with evening than morning, noon, or night, and its bright white and yellow colors suggest light and sunshine. It does have a thick beak, as do many finches, weaverbirds, cardinals, and tanagers.

Like most finches, evening grosbeaks gather into flocks in the fall and winter. I own a fond memory of a flock of them on a fall day over fifty years ago. The leaves were down, and the birds were feeding on the seeds hanging from a white ash tree. With their bright colors, compact body shapes, thick bills, and gregarious habits, they seemed more like parrots than finches. The picture of them in the ash tree stuck in my mind. That may have been my first encounter with them. But evening grosbeaks are now common winter visitors to feeders supplied with sunflower seeds. In one of my still-favorite bird-identification books (*Audubon Bird Guide,* 1946), the author speculates that winter is a critical period for the grosbeaks because of low food supplies, and that "if enough winter bird-feeding stations are established to free the evening grosbeak from dependence on natural foods its populations might increase greatly in years to come."

Sixty-eight years have passed since Pough's hopeful pronouncement, and there are probably as many or more bird-feeding stations as in his time, but I doubt that the number of these birds has increased much. Like pine grosbeaks, redpolls, pine siskins, red crossbills, and white-winged crossbills, they may be in great abundance one year and then absent for many years. They are, like other seed-eating finches, wanderers that largely depend on the vagaries of wild seed crops. Whenever I hear their metallic ringing calls I take notice — as I did in the spring of 2011.

During the winter of 2010–2011, I did not hear or see evening grosbeaks in my western Maine woods. I was not surprised, since it was one of the rare years when there were, to my knowledge, no trees offering seeds in the fall and winter. The sugar maple had not flowered that spring, and there were no seeds of pine, hemlock, larch, red spruce, white spruce, or balsam fir. Nor did I see any ash, striped maple, or mountain maple seeds, or any birch seed of the three common species (yellow, gray, and white). In short, there was an unusual paucity of tree seeds. Even the red-breasted nuthatches that have in my memory "always" been there were absent that winter. Might they or any other seed eaters return to nest in the spring?

Near the end of the first week of May 2011, the trees were still leafless. But the red maples, which flower every spring, were nearly finished blooming. Pale yellow patches of the mass-flowering sugar maples, which do not flower every year, dotted the hillsides against the gray-brown of leafless deciduous trees and the dark green conifers. I was cheered to hear, for the first time since the previous year, the calls of a red-breasted nuthatch. The summer birds were flooding in, and the songs of ovenbirds, Nashville warblers, yellowthroat warblers, purple finches, black-throated green warblers, and solitary vireos sounded all around. Yellow-bellied sapsuckers, northern flickers, and pileated, downy, and hairy woodpeckers were drumming here and there. Four tree swallows circled high over the clearing at dawn, and two of them stayed and carried the first dry grass into a bird box. But then on May 7, to my surprise, three evening grosbeaks landed on the sugar maples by my cabin.

At dawn four days later, on May 11, temperatures dropped to near 40°F under an overcast sky. It was not a great day to see birds, but still, as I opened the cabin door I again heard the distinctive bell-like calls of evening grosbeaks, and there, right in front of me, a flock of seventeen was perched on a young quaking aspen tree on the other side of my clearing. The adult males are clad in bright white, yellow, and brown plumage, with females and immature males in muted grayish yellow. But even more startling to me was the color of the birds' bills. I looked again to be sure. Yes, they were pistachio green, unlike the ivory of the evening grosbeaks that I had seen before or those in the latest bird guide I had received as a gift. The green shone prettily against the males' luminescent yellow feathers.

I was captivated by the sight, and noticed that these grosbeaks seemed to be picking at leaves — I had the impression that they were perhaps eating them, but I doubted my own eyes. Grosbeaks are seed eaters. Their thick bills evolved to be able to crack cherry pits. After I watched them in astonishment for some minutes, they all left, as if at a signal, in their rapid parrotlike flight. I climbed up and picked a twig from where they had been, to see if I could tell whether they had really been leaf grazing. The twig seemed sparsely leafed out, but then I counted sixty still-attached fresh leaf petioles without their leaf blades, which must indeed have been snipped off. The pistachio-colored bills had not been a hallucination either. As I later learned, evening grosbeaks are one of the few if not the only songbirds that change beak color in the nesting season, as some birds change into nuptial garb by molting their old feathers and growing new.

Evening grosbeaks are at home and nest in the boreal zone of fir and spruce. I did not see the flock of seventeen again. Perhaps they traveled to Canada and northern Maine, where there are endless spruce-fir woods, and where I had seen them nest in June in the early 1960s, when as a student I was working in the Allagash in a lumbering camp.

Ten days later, on May 21, some evening grosbeaks were still around: a flock of seven flew over. At this time the beech leaves had fully unfurled, the maples were almost leafed out, and blue violets had erupted on the forest floor along with white and purple trilliums. I saw the head of an incubating broad-winged hawk as she peered down over the nest edge. The leaf-gleaning birds such as red-eyed vireos and scarlet tanagers were back, and the phoebe had started to lay eggs.

In another week, while I was watching nesting tree swallows, I was again distracted by the loud ringing calls of an evening grosbeak. I judged the sound came from the hardwood forest nearby, and I heard it again from the same direction several hours later. Odd, I thought, because if there is one thing these birds do consistently it is to move around a lot. I had not heard any for days.

When I went to check I found a male grosbeak perched in the top of a huge pine tree among the maples, ash, and poplars. A female appeared and landed briefly on a branch next to him. They exchanged soft whispered calls. When she flew off he followed closely behind her. I sat quietly in the woods where they seemed to have vanished, and soon again heard their soft, almost whispering calls — and then saw the pair less than five meters above me. Their sweet-talking indicated that they were a mated pair, and given the season, they should and could be nesting. They next flew into a dense young red spruce growing among the hardwoods, and from there she flew to the same big pine where I had earlier seen the male, and he followed right behind her. Again he made the loud metallic clanging calls. She returned to a spot near me, and together they flew back to the same red spruce. When she picked at a dry dead twig, I ran back to the pine to await them, and as expected they revisited it. I saw her fly to their just-begun nest at the tip of a pine branch some thirty meters above the ground.

In contrast to the previous year, the pine tree was laden with young cones. Already the balsam firs, red and white spruces, white ash, and sugar maples were also fruited. It promised to be an excellent seed year.

Much was new. I saw grosbeaks eat leaves, change beak color, build a nest, and act as a couple. I was used to seeing the grosbeaks in flocks almost every winter, but unlike the other finches, they wear the same-colored feather garb year-round. Might the green color of their bills at this time be attributable to their food? Might it signal that they were ready to nest? Or is it related to sexual selection? But if so, then only one sex should have the specific bill coloration. Contrasts pique interest, in bird and man.

17

~

Audience to a Woodcock

THE AMERICAN WOODCOCK IS A UNIQUE CLOWNISH-LOOK-
ing bird classified with the sandpipers and shorebirds in the family Scolpacidae — but it's a shorebird that has nothing to do with
shores. It walks on stubby legs, has a bill a third of its body length,
and has eyes placed toward the back of its head. At dusk on a
cool evening in April, the male flies over the trees into a clearing
and, on fluttering wings, settles onto the damp matted grass and
leaves where the snow has recently melted. He makes a series of
strange-sounding soft hiccups, and then moves forward in short
jerky motions interspersed with loud buzzy *peent*s. As it gets dark
he rockets up into the air like a giant hummingbird, his wings
making a whistling sound. At first he flies up at an angle of about
thirty degrees, rising ever higher over the forest. After reaching
an altitude of a hundred meters or so, he starts to circle and rises
so high that he is merely a speck in the sky, then even the speck
disappears. You hear him only faintly, if there are no competing

sounds, or not at all. But then you begin to detect a rapid series of high-pitched chirps as he starts to plummet in spirals and zigzags. Near treetop level he breaks his descent with a loud fluttering of wings, and he lands at almost the exact spot from which he took off about a minute earlier.

Long before I knew a thing about woodcocks I had seen and heard their sky dance. As a boy I once stayed all night in a field near our farm in Maine after hearing a woodcock there in the evening. At dawn, after snow had covered my sleeping bag, I heard it again. The bird's music was to me far more engaging than what played on the radio. When about half a century later I made a clearing in the woods next to my Maine cabin, the one avian musician I had in mind was the woodcock. Every year I look forward to his spring ritual, and I have not missed the April premiere for decades.

Since the woodcock is a popular game bird, its habits and habitat have been and still are of great interest. This bird is perhaps one of the most-studied in North America. We now know that, any spring in a given spot, the sky dance is likely to be performed by the same individual as in previous years. But there is also a good chance that he has been replaced. Ownership of a launching plot is fought for, and territorial success depends on a male's dominance over other males, who are under pressure to perform, as during a courtship period a female may visit up to three different males in a single evening. Since males probably abandon any display area where no females visit them, their display areas are apt to be near habitat suitable for rearing young.

The woodcock has been experiencing a downward plunge in population for thirty years because of forest regrowth. Oddly, although this bird lives on the ground under a cover of thick woodland trees, for the species to reproduce the male requires a sizeable clearing as a launch pad for his aerial mating dance. (I do my part. I have laboriously used a chainsaw, an axe, and brush cutters to hold back the constantly incoming tide of forest to maintain a cleared space in my woods.) And the woodcock also faces other challenges. Evolution of the males' conspicuous display has been fueled by competition to attract receptive females — but the display that attracts females can also lure predators.

Food may be a problem as well. Woodcocks feed on earthworms and reach their food by probing with their long bill, which has a hook on its tip for holding a worm. (Judging from the numerous poke holes left on a patch of mud where a woodcock has foraged, it seems likely that the bird pokes into the mud at random, and one wonders how it manages to find any worm at all.) Woodcocks likely gamble their lives every spring by arriving from the south when the ground is apt to be frozen hard or blanketed in deep snow, conditions that prevent them from reaching the earthworms they feed on.

In 2011 I saw my first woodcock near the usual time and usual conditions, on March 26. Temperatures had been near −12°C for a few days, but were rising as I was driving to my Maine cabin from Vermont. At 6:30 p.m., when the light was fading, I spotted a woodcock along the roadside. It was not moving. I turned my pickup around and drove back, hoping to pick up the bird if it was roadkill. But by the time I got back it had walked across the road and up a snowbank, and I saw it fly off in typical whistling flight. The ground was frozen solid. At the cabin all was still under deep snow, although some steep south-facing slopes had bare spots. It was not until almost a month later, on April 21, when there had been significant snowmelt in my clearing, that I saw a sign. I saw the white woodcock droppings (sometimes referred to as paint spots) at a damp depression from meltwater at the lower edge of my clearing. A woodcock had come. And at 7:55 that evening, I heard his *peent*s and saw him launch into his sky dance. He had made it home again! I was elated to once again partake of his presence.

The year before this, on May 5, friends from Sterling College and I had found two just-hatched woodcock chicks near this display ground. The downy chicks were revealed after we almost stepped on the mother, who was as camouflaged as her babies. They had been under her, and after she flushed they flattened themselves into the loose leaves. I sketched them, to preserve what they looked like while they were trying to look like nothing at all — or like everything on a leaf-littered forest floor.

This year (2012), on April 25, I went to the same area to look for a possible nest. The nests, which female woodcocks make for their three or four camouflaged eggs, look like mere depressions or scrapes on the ground.

As soon as I got there I flushed a woodcock. To my great surprise, a second one remained at almost the same spot from which the other had left. She (the sex was my assumption; males and females look identical) walked slowly away from me, taking a few steps, stopping, and then gently rocking her body forward and back. As I advanced a few steps to take a photograph, she repeated the curious rocking motions and walked farther ahead of me. She made no attempt to hide, and I wondered why she rocked, making herself conspicuous. Perhaps to lead me away from her nest or babies? Maybe, but I recalled surprising a woodcock with young on a road some years earlier. She had made a squealing noise and then, like many ground-nesting birds, having caught my attention, had feigned a broken wing to entice me to follow her.

This woodcock's rocking walk was different, but I had no doubt that it too was meant to get my attention. I continued following as she walked and rocked ahead of me. Finally she flew up, but went only about ten meters before landing back on the ground among the stems of a dense winterberry bush. What would she do now? Maybe go back to a nest? I withdrew thirty meters or so to wait and watch. As soon as I withdrew several meters she stopped walking and after a few moments also stopped rocking. Fifteen minutes later she still hadn't budged. I walked back to the bush, and she moved to the opposite side of it, then walked on as before while I slowly followed. Soon, however, I turned and walked away, to make her think she had indeed outwitted me.

I happened to walk there again two weeks later, after having watched the nightly sky dances the evening before. This time, near noon, "the" woodcock was back at the same spot. Again she walked in front of me. But this time she stopped and flared her stubby tail feathers to show flashy white feathers that would normally be hidden from view. Now feeling even more convinced that she was trying to lead me away, I played along and kept following to see where we might end up. As before, she stopped whenever I stopped, then gently rocked her body. But this time instead of forward and back she rocked up and down, as if she were mounted on a light spring. Thinking that either eggs or babies were surely nearby, I searched for ten minutes, but found nothing.

If the woodcock was indeed a female, and if she had eggs or small young, she would return to the nest. So I climbed high into a maple tree that was behind other trees and far enough away that I would not be seen or at least not be seen as a threat. But I was close enough to be able to see with my binoculars. If she had young she would have to go back, or at least call to them to come out of their hiding places. I had a pleasant time in the tree. I saw a blue butterfly and also a fritillary butterfly below me fluttering along the ground. But the woodcock remained at the same place, in the wet depression at the edge of the field near the landing platform where the male had *peent*ed and launched into his sky dance.

I would again have let the question drop, assuming that the bird was a female leading me away from nest and eggs, and that I had been simply inept at finding them. But then a friend in Trenton, Maine, sent me a video he had taken of a woodcock doing the same thing, a month earlier, and in a driveway. This caught my attention, because it convincingly disproved my hypothesis: there was no way that his woodcock would have had a nest at that time, immediately after returning from migration. I now had an interesting problem. The woodcock, we recall, is one of the most-studied birds. So, not seeing the behavior again, I went to the Internet to look for an explanation of my woodcock's puzzling behavior.

I found videos that showed woodcock doing precisely what I had seen — a behavior often described as the "swaying dance walk" — but that more or less poked fun at woodcocks as either simply stupid or uncannily endowed with almost magical powers. Biologists have long noted the behavior, and have proposed imaginative hypotheses to explain it. The first, suggested by O. S. Pettingill (1936), was "fear and suspicion." Fear and suspicion may or may not be involved, but how would anyone prove it, and so what? "Reducing detection by minimizing shadows" was another hypothesis, offered by C. B. Worth (1976). But it is a stretch to think a woodcock acting as though suspended from a yo-yo reduces shadows, especially when ducking down and hugging the ground so clearly does. An alternative hypothesis, the most recent I found, was proposed in 1982: W. M. Marshall posited that the swaying dance walk was the way woodcocks forage for angleworms. This one paints a picture of woodcocks as truly ingenious. It goes this way: when earthworms detect vibrations such as those created by a bobbing woodcock, they squirm, and the woodcock detects the movement with its feet and probes for the squirming worm. Not everyone agrees that worms would make their presence known to woodcocks by squirming. Some claim instead that worms squirm their way out of the ground to the surface as a defense against tunneling moles that hunt them. So woodcocks' rocking is really a mechanism to mimic the presence of moles so as to induce the worms to surface. Then, presto, the woodcocks get their meals. The problem is that, to my knowledge, none of these steps, never mind this sequence of them, has ever been observed, nor has anyone ever seen a woodcock actually capture a worm by this method. I don't buy the hypothesis, simply by going by facts. All the rocking I saw was gentle and unlikely to have caused vibrations of the earth. Nor would a woodcock go worm hunting in the presence of a potential threat: instead, its impulse would be to fly away. And during worm hunting a woodcock would not stop hunting the moment a pursuer stopped, and proceed the moment it pursued again.

After replaying in my head what I'd seen of the swaying dance walk and re-watching my friend's video of it recorded in early April when the ground was still frozen, I realized that this rocking-swaying walk was indeed intriguing, and that although proving what it did might be difficult, there was plenty of proof of what it could not do. For example, performing it on ice or on gravel drive-ways (as in several videos) could not yield earthworms. Also, when I sampled the icy ground where I had seen this display, the few worms I dug up were still unreactive and unable to crawl. Thus the swaying dance walk makes no sense as a foraging technique, but more sense as a distraction display, like the flashing of the white tail flag by a deer, the rattle of a rattlesnake, or the whistle of the wings of a dove when startled by potential predators.

The woodcock ground display, I now felt sure, alerts predators that they have been seen. Once a predator spots the bird, the pecu-liar placement of the woodcock's eyes gives it the impression that the bird is watching it all the time. (Predators are acutely aware of eyes in assessing their chances of capturing prey.) Meanwhile, the rocking walk reinforces the signal that the predator has lost the advantage of surprise, and that continuing to hunt this prey would be a waste of time. If the display succeeds, the woodcock does not need to fly away, whether from a patch of cover, its favor-ite launch site for aerial displays, or perhaps just a good spot for finding worms.

The woodcock does its fabulous sky dance for one audience and its rocking walk dance for another, but both in order to be heard or seen. And so I too, prompted by a hypothesis, have been an audi-ence to my woodcock. I was back at the cabin full time the next spring, but saw the woodcock only briefly. There were several abbreviated sky dances; then there was silence, except for the frogs. Just at the time for the sky dances, the vernal pool I had dug at the edge of the clearing was beginning its annual mating choruses of wood frogs and spring peepers. About five years earlier I had heard the first wood frog there, but each year there were more: in 2014 more than two hundred females each laid a clutch of about two hundred eggs. The sound of nearly as many males was deafening, especially when combined with the even greater din of spring peepers. A woodcock male encountering such noise pollution would be unlikely to attract an audience, and might leave. For the hypothesis that birds respond to ambient noise there is precedent: birds living in noisy cities sing louder than those in the quiet countryside, suggesting that they calibrate their display with respect to their audience.

I have been an audience to a woodcock every spring since I heard my first one as a boy at my family's farm. I especially recall the woodcock that came to a patch of snowmelt near an alder thicket in our back field. I saw it perform one night, and to hear it again, and again and again, required my getting a grandstand seat, which was a tent pitched as close as the bird allowed to the small scrub pine where it landed after each performance. I watched and learned, with no theory to prove. And now, sixty years later, I'm still learning by being an audience to a woodcock, and so can anyone learn by watching a starling, a sapsucker, a flicker, or a house sparrow — one wild bird at a time.

Acknowledgments

———

BIRD WATCHING AND WRITING ARE BEST DONE SOLITARILY.
However, making a book about the two may involve, among other
things, suggestions and corrections, nudgings and judgings, and
companionship of friends and kindred souls who critique and act
as real or imagined sounding boards. In this case, they included (in
no particular order) Alan Burger, Andrea Lawrence, Paul Spitzer,
Lillian Reade, John Alcock, Margaret McVey, Greg Fell, Dean
Leslie, Charles Sewall, Glenn Booma, Albert Reingewirtz, Andrea
Lawrence, John Marzluff, Peter Miller, Duane and Nancy Leavitt,
Lance Lichtensteiger, Alexandra and Garrett Conover, Joel Babb,
and Lynn Jennings. Of these I acknowledge especially John Alcock for reading a draft of the manuscript and making valuable
technical suggestions. I also thank my agent, Sandra Dijkstra, and
Lynn Jennings, who have often nudged me in the right direction
and helped me keep on course. I give sincere thanks to the editors
and production staff of the publisher, Houghton Mifflin Harcourt,
in this case specifically Camille Smith and Deanne Urmy, whose
critical watchful eyes were in many cases more perceptive than

mine. Last but not least, I have profited greatly from the generous support of Craig Neff and Pamelia Markwood, of the Naturalist's Notebook of Seal Harbor, Maine, for digitizing, archiving, and storing almost all of my illustrations and making the appropriate ones conveniently available to be used for this publication.

ACKNOWLEDGMENTS

Further Reading

———

FLICKERS IN THE HOUSE

Bent, Arthur C. 1939. *Life Histories of North American Woodpeckers.* Washington, DC: Government Printing Office. 342 pp.

Kilham, L. 1959. Early reproductive behavior of flickers. *Wilson Bulletin* 71:323–336.

———. 1983. *Life History Studies of Woodpeckers of Eastern North America.* Cambridge, MA: Nuttall Ornithological Club. 240 pp.

Sherman, A. R. 1910. At the Sign of the Northern Flicker. *Wilson Bulletin* 22:135–171.

A QUINTET OF CROWS

Caffree, C. 1992. Female-biased delayed dispersal and helping in American crows. *The Auk* 109(3):609–619.

Heinrich, B. 1999. *Mind of the Raven: Investigations and Adventures with Wolf-Birds.* New York: Harper-Collins. 380 pp.

Kilham, L. 1984. Cooperative breeding of American crows. *Journal of Field Ornithology* 55(3):349–356.

——. 1985. Behavior of American crows in the early part of the breeding cycle. *Florida Field Naturalist* 13(2):25–48.

Marzluff, J. M., and T. Angell. 2005. *In the Company of Crows and Ravens.* New Haven, CT: Yale University Press. 384 pp.

Marzluff, J. M., J. Walls, H. N. Cornell, J. Withey, and D. P. Craig. 2010. Lasting recognition of threatening people by wild American crows. *Animal Behaviour* 79:699–707.

Verbeek, N. A., and C. Caffrey. 2002. American Crow (*Corvus brachyrhynchos*). *The Birds of North America Online,* ed. A. Poole. Ithaca: Cornell Laboratory of Ornithology. http://bna.birds.cornell.edu/BNA/

GETTING TO KNOW A STARLING

Baptista, L. F., and L. Petrinovich. 1984. Social interaction, sensitive periods, and song template hypothesis in the white-crowned sparrow. *Animal Behaviour* 36:1753–64.

Gentner, T. Q., K. M. Fenn, D. Margoliash, and H. C. Nusbaum. 2006. Recursive syntactic pattern learning by songbirds. *Nature* 440:1204–07.

Murmuration of Starlings. See http://www.youtube.com/embed/88UVJpQGi88.

West, M. J., A. N. Stroud, and A. P. King. 1983. Mimicry of the human voice by European starlings: The role of social interaction. *Wilson Bulletin* 95:635–640.

West, M. J., and A. P. King. 1990. Mozart's Starling. *American Scientist* 78:106–114.

WOODPECKER WITH A DRUM

Bent, A. C. 1939. Life Histories of North American Woodpeckers. *Bulletin of the U.S. National Museum* 174:1–1322. Reprinted by Dover Publications, New York, 1964.

Daily, G. C., P. R. Ehrlich, and N. M. Haddad. 1993. Double keystone bird in a keystone species mix. *Proceedings of the National Academy of Sciences USA* 90:592–594.

Kilham, L. 1983. *Life History Studies of Woodpeckers of Eastern North America.* Nuttall Ornithological Club no. 20:1–240.

Walters, E. L., E. H. Miller, and P. E. Lowther. 2002. Yellow-bellied Sapsucker (*Sphyrapicus varius*). *The Birds of North America,* no. 662, ed. A. Poole and F. Gill. Philadelphia: The Birds of North America, Inc.

Barred Owl Talking

Angell, T. 2015. *The House of Owls.* New Haven, CT: Yale University Press. 203 pp.

Bent, A. C. 1938. *Life Histories of North American Birds of Prey,* part 2. U.S. National Museum Bulletin, no. 170.

Eckert, A. W. 1974. *The Owls of North America.* New York: Doubleday.

Freeman, P. L. 2000. Identification of individual barred owls using spectrogram analysis and auditory cues. *Journal of Raptor Research* 34:85–92.

Heinrich, B. 1987. *One Man's Owl.* Princeton, NJ: Princeton University Press. 224 pp.

Klatt, P. H., and G. Richison. 1993. The duetting behavior of eastern screech owls. *Wilson Bulletin* 105(3):483–489.

Odum, K. L., and D. J. Merrill. 2010a. A quantitative description of the vocalizations and vocal activity of the barred owl. *Condor* 112:549–560.

——. 2010b. Vocal duets in nonpasserines: An examination of territorial defense and neighborhood-stranger discrimination in a neighborhood of barred owls. *Behaviour* 147:619–639.

——. 2012. Inconsistent geographic variation in the calls and duets of barred owls (*Strix varia*) across an area of genetic introgression. *The Auk* 129(3):387–398.

Hawk Tablecloths

Berger, S., R. Disko, and H. Gwinner. 2003. Bacteria in starling nests. *Journal of Ornithology* 144:317–322.

Brouwer, L., and J. Komdeur. 2004. Green nesting material has a function in mate attraction in the European starling. *Animal Behaviour* 67:539–548.

Clark, L., and J. R. Mason. 1985. Use of nest material as insecticidal and anti-pathogenic agents by the European starling. *Oecologia* 67:169–176.

Gracelin, D. H. S., A. J. Britto, and P. B. J. R. Kumar. 2012. Antibacterial screening of a few medicinal ferns against antibiotic resistant phytopathogens. *International Journal of Pharmaceutical Sciences and Research* 3:868–873.

Gwinner, H., and S. Berger. 2005. European starlings: Nestling condition, parasites, and green nesting material during the breeding season. *Journal of Ornithology* 146:365–371.

Gwinner, H., M. Oltrogge, L. Trost, and U. Nienaber. 2000. Green plants in starling nests: Effects on nestlings. *Animal Behaviour* 59:301–309.

Heinrich, B. 2010. *The Nesting Season: Cuckoos, Cuckolds, and the Invention of Monogamy.* Cambridge, MA: Harvard University Press. 352 pp.

Heinrich, B. 2013. Why does a hawk build with green nesting material? *Northeastern Naturalist* 20(2):209–218.

Hoffman, D. 2003. *Medical Herbalism: Principles and Practice.* Rochester, VT: Healing Arts Press. 588 pp.

Lyons, D. M., K. Titus, and J. A. Mosher. 1986. Sprig delivery by broad-winged hawks. *Wilson Bulletin* 98:469.

Matray, P. F. 1974. Broad-winged hawk nesting ecology. *The Auk* 91:307–324.

Orians, G. F., and F. Kuhlman. 1956. The red-tailed hawk and great horned owl populations in Wisconsin. *Condor* 58:371–385.

Rodgers, J. A., Jr., A. S. Wenner, and S. T. Schwikert. 1988. The use and function of green nest material by wood storks. *Wilson Bulletin* 100:411–423.

Rosenfield, R. N. 1982. Sprig collection by a broad-winged hawk. *Raptor Research* 16:63.

Srivastava, K. 2007. Importance of ferns in human medicine. *Ethnobotanical Leaflets* 11:231–234.

Welty, J. C. 1962. *The Life of Birds.* Philadelphia: W. B. Saunders. 720 pp.

Wimberger, P. H. 1984. The use of green plant material in bird nests to avoid ectoparasites. *The Auk* 101:615–618.

Vireo Birth Control

Alcock, J. 2005. *Animal Behavior: An Evolutionary Approach,* 8th ed. See ch. 12, The evolution of parental favoritism, pp. 426–435.

Husby, M. 1986. On the adaptive value of brood reduction in birds: Experiments with magpies, *Pica pica. Journal of Animal Ecology* 55:75–83.

James, R. D. 1998. Blue-headed Vireo (*Vireo solitarius*). *The Birds of North America,* no. 379., ed. A. Poole and F. Gill. Philadelphia: Academy of Natural Sciences.

Mock, D. W. 1984. Siblicide aggression and resource monopolization in birds. *Science* 225:731–733.

Mock, D. W., H. Drummond, and C. H. Stinson, 1990. Avian siblicide. *American Scientist* 78:438–449.

O'Connor, R. J. 1978. Brood reduction in birds: Selection for fratricide, infanticide and suicide? *Animal Behaviour* 26:790–796.

——. 1979. Egg weights and brood reduction in the European swift (*Apus apus*). *Condor* 81:133–145.

Nuthatch Homemaking

Ghalambor, C. K., and T. H. Martin. 1999. Red-breasted Nuthatch (*Sitta canadensis*). *The Birds of North America Online,* ed. A. Poole. Ithaca, NY: Cornell Laboratory of Ornithology. http://bna.birds.cornell.edu/BNA/.

Blue Jays in Touch

Johnson, W. C., and T. Webb. 1989. The role of blue jays (*Cyanocitta cristata*) in the postglacial dispersal of fagaceous trees in North America. *Journal of Biogeography* 16:561–571.

Jones, T. B., and A. C. Kamil. 1973. Tool-making and tool-use in the northern blue jay. *Science* 180:1076–78.

Racine, R. N., and N. S. Thompson. 1983. Social organization of wintering blue jays. *Behaviour* 87:237–255.

Stewart, P. A. 1982. Migration of blue jays in eastern North America. *North American Bird Bander* 7:107–112.

Tarvin, K. A., and G. E. Woolfenden. 1999. Blue Jay (*Cyanocitta cristata*). *The Birds of North America*, no. 469, ed. A. Poole and F. Gill. Philadelphia: Academy of Natural Sciences.

Chickadees in Winter

Barnea, A., and N. Nottebohm. 1994. Seasonal recruitment of hippocampal neurons in adult free-ranging black-capped chickadees. *Proceedings of the National Academy of Sciences USA* 91:11214–221.

Heinrich, B., and R. Bell. 1995. Winter food of a small insectivorous bird, the golden-crowned kinglet. *Wilson Bulletin* 107(3):558–561.

Heinrich, B., and S. L. Collins. 1983. Caterpillar leaf damage, and the game of hide-and-seek with birds. *Ecology* 64(3):592-602.

Nottebohm, F. 1980. Testosterone triggers growth of brain vocal control nucleus in adult female canaries. *Brain Research* 189:429–436.

——. 1981. A brain for all seasons: Cyclical anatomical changes in song control nucleus of the canary brain. *Science* 214:1368–70.

——. 1989. From birdsong to neurogenesis. *Scientific American* 260:74–79.

Sherry, D. F., and J. S. Hooshooly. 2009. The seasonal hippocampus of food-storing birds. *Behavioral Processes* 80:334–338.

Smith, S. M. 1991. *The Black-capped Chickadee: Behavioral Ecology and Natural History.* Ithaca, NY, and London: Comstock Publishing Associates of Cornell University Press. 362 pp.

Redpolls Tunneling in Snow

Cade, T. J. 1953. Sub-nival feeding of the redpoll in interior Alaska: A possible adaptation to the northern winter. *Condor* 55:43–44.

Clement, R. C. 1968. Common redpoll. In *Life Histories of North American Cardinals, Grosbeaks, Buntings, Finches, Sparrows, and Allies,* ed. A. C. Bent and O. L. Austin. Washington, DC: Smithsonian Institution Press.

Collins, J. E., and J. M. C. Peterson. 2003. Snow burrowing by common redpolls (*Carduelis flammea*). *The Kingbird* 53(1):13–22.

Furness, G. 1987. Common redpolls excavating snow burrows and snow bathing. *The Kingbird,* Spring, pp. 74–75.

Guntert, M., D. Hay, and R. P. Balda. 1988. Communal roosting in the pygmy nuthatch: A winter survival strategy. *Proceedings of the International Ornithological Congress* 19:1964–72.

Heinrich, B. 2014. Redpoll snow bathing: Observations and hypothesis. *Northeastern Naturalist* 21(4):N45–N52.

Heinrich, B., and R. Smolker. 1998. Play in common ravens (*Corvus corax*). In *Animal Play: Evolutionary, Comparative, and Ecological Perspectives,* ed. M. Beckoff and J. A. Byers, pp. 27–44. Cambridge, UK: Cambridge University Press.

Knox, A. G., and P. E. Lowther. 2000. Common redpoll (*Carduelis flammea*). *The Birds of North America Online,* ed. A. Poole and G. Gill. Ithaca, NY: Cornell Laboratory of Ornithology. http://bna.birds.cornell.edu/BNA/.

Korhonen, K. 1981. Temperature in the nocturnal shelters of the redpoll (*Acanthis flammea* L.) and the Siberian tit (*Parus cinctus* Budd.) in winter. *Annales Zoologici Fennici,* pp. 165–168.

Meltofte, K. 1983. Arrival and pre-nesting period of the snow bunting *Plectophenax nivalis* in East Greenland. *Polar Research* 1:185–198.

Novikov, G. A. 1972. The use of under-snow refuges among small birds of the sparrow family. *Aquilo Serie Zoologica* 13:95–97.

Palmer, R. S. 1949. Maine Birds. *Bulletin of the Museum of Comparative Zoology* 102.

Sulkava, S. 1968. On small birds spending the night in the snow. *Aquilo Serie Zoologica* 7:33–37.

TRACKING GROUSE IN WINTER

Bump, G. R., R. W. Darrow, F. C. Edminster, and W. F. Crissey. 1947. *The Ruffed Grouse: Life History, Propagation, and Management.* Buffalo: New York State Conservation Department. 915 pp.

Heinrich, B. 2003. *The Winter World: The Ingenuity of Animal Survival.* New York: HarperCollins, p. 347.

——. 2004. Overnighting of golden-crowned kinglets in winter. *Wilson Bulletin* 115:123–124.

Page, R. E., and A. T. Bergerud. 1988. A genetic explanation for the ten-year cycles in grouse. In *Adaptive Strategies and Population Ecology of Northern Grouse,* ed A. T. Bergerud and M. W. Gratson. Minneapolis: University of Minnesota Press.

Whitaker, D. M., and D. F. Stauffer. 2003. Night roost selection during winter by ruffed grouse in the central Appalachians. *Southern Naturalist* 2(3):377–392.

CRESTED FLYCATCHER'S NEST HELPERS

Alcock, J. 2005. *Animal Behavior: An Evolutionary Approach,* 8th ed., pp. 405–435. Sunderland, MA: Sinauer Associates.

Clutton-Brock, T. H. 1991. *The Evolution of Parental Care.* Princeton, NJ: Princeton University Press.

Davies, N. B. 2000. *Cowbirds and Other Cheats.* London: T. and A. D. Poyser.

RED-WINGED BLACKBIRDS RETURNING

Orians, G. H. 1980. *Marsh-nesting Blackbirds*. Princeton, NJ: Princeton University Press.

Searcy, W. A., and K. Yasukawa. 1995. *Polygyny and Sexual Selection in Red-winged Blackbirds*. Princeton, NJ: Princeton University Press.

PHOEBE SEASONS

Heinrich, B. 2000. Phoebe diary. *Natural History* 109(4):14–15.

Olson, Roberta, and New York Historical Society. 2012. *Audubon's Aviary: The Original Watercolors for The Birds of America*. New York: Skira Rizzoli.

Pough, Richard H. 1946. *Audubon Bird Guide*. New York: Doubleday.

EVENING GROSBEAKS

Gillihan, S. W., and B. Byers. 2001. Evening grosbeak (*Coccothraustes vespertinus*). *The Birds of North America Online*, ed. A. Poole. Ithaca, NY: Cornell Laboratory of Ornithology. http://bna.birds.cornell.edu/BNA/

AUDIENCE TO A WOODCOCK

Longcore, J. R., D. G. McAuley, G. F. Sepic, and G. W. Pendleton. 1996. *Canadian Journal of Zoology* 74:2046–54.

Marshall, W. M. 1982. Does the American woodcock bob or rock — and why? *The Auk* 99:791.

McAuley, D. G., J. R. Longcore, and G. F. Sepic. 1993. Behavior of radio-marked breeding American woodcocks. *Proceedings of the Eighth American Woodcock Symposium*, pp. 116–125.

Mendell, H. L., and C. M. Aldous. 1948. *The Ecology and Management of the American Woodcock*. Orono: Maine Cooperative Wildlife Research Unit, University of Maine. 201 pp.

Nemeth, E., and H. Brumm. 2009. Blackbirds sing higher-pitched songs in cities: Adaptation to habitat acoustics or side-effect of urbanization. *Animal Behaviour* 78(3):637–641.

Pettingill, O. S., Jr. 1936. The American woodcock (*Philohela minor*). *Memoirs of the Boston Society of Natural History* 9:169–391.

Sheldon, W. G. 1967. *The Book of the American Woodcock.* Amherst: University of Massachusetts Press.

Worth, C. B. 1976. Body-bobbing woodcocks. *The Auk* 93:374–375.

Latin names of the highlighted or commonly cited birds

———

American Crow	*Corvus brachyrhynchos*
American Goldfinch	*Carduelis tristis*
American Robin	*Turdus migratorius*
Bank Swallow	*Riparia riparia*
Barred Owl	*Strix varia*
Black-capped Chickadee	*Poecile atricapilla*
Blue-headed Vireo	*Vireo solitaries*
Blue Jay	*Cyanocitta cristata*
Broad-winged Hawk	*Buteo platypterus*
Canada Goose	*Branta canadensis*
Cliff Swallow	*Petrochelidon pyrrhonota*
Common Grackle	*Quiscalus quiscula*
Common Raven	*Corvus corax*
Common Redpoll	*Carduelis flammea*
Eastern Phoebe	*Sayornis phoebe*
European Starling	*Sturnus vulgaris*

Evening Grosbeak	*Coccothraustes vespertinus*
Great Crested Flycatcher	*Myiarchus crinitus*
Great Horned Owl	*Bubo virginianus*
Mourning Dove	*Zenaida macroura*
Northern Flicker	*Colaptes auratus*
Northern Saw-whet Owl	*Aegolius acadicus*
Pine Siskin	*Carduelis pinus*
Red-breasted Nuthatch	*Sitta canadensis*
Red-tailed Hawk	*Buteo jamaicensis*
Red-winged Blackbird	*Agelaius phoeniceus*
Ruby-throated Hummingbird	*Archilochus colubris*
Ruffed Grouse	*Bonasa umbellus*
Sharp-shinned Hawk	*Accipiter striatus*
Tree Swallow	*Tachycincta bicolor*
Wild Turkey	*Meleagris gallopavo*
Yellow-bellied Sapsucker	*Sphyrapicus nuchalis*
Yellow-rumped Warbler	*Dendroica coronata*

Index

———

Page numbers in *italics* indicate illustrations.